OUT OF THE EARTH

OUT OF
THE EARTH

By Louis Bromfield

New York

HARPER & BROTHERS PUBLISHERS

For permission to reprint selections from the following books acknowledgment is made to Dodd, Mead & Company for *The Insect World of J. Henri Fabre* by Edwin Way Teale. Copyright, 1917, 1918, by Dodd, Mead & Company, Inc.; Little, Brown & Company for *Our Plundered Planet* by Fairfield Osborn; Harcourt, Brace & Company for *Give Us This Day* by Clare Leighton; Charles Scribner's Sons for *Cry, the Beloved Country* by Alan Paton.

DEDICATION

For all the friends of good agriculture and animal husbandry throughout the nation and especially for the friends in the Great Southwest
and at Wichita Falls, Texas, who helped to open up the whole of
that vast, rich region to the author and enabled him and Bob Huge
to set up a second Malabar Farm in the valley of the Wichita River.

The author is especially grateful to *The Farm Quarterly* and *The Country Gentleman* for permission to use certain photographs included in the book, and to Joe Munroe and Robert Hadley of Harry Ferguson, Inc., for the excellent photographs which come from the record both have made of the operations at Malabar for some years past. Gratitude is also expressed to others who have contributed from time to time photographs made at Malabar in the interest of agriculture in general.

Contents

*Thirty-two pages of photographs, printed as a separate
section, follow page 114*

The State is like a tree. The roots are agriculture, the trunk is the population, the branches are industry, the leaves are commerce and the arts; it is from the roots that the tree draws the nourishing sap . . . and it is to the roots that a remedy must be applied if the tree is not to perish.

—Victor, Marquis de Mirabeau
(written at the beginning of
the eighteenth century)

Foreword

THIS book has been written largely at the request of the thousands of practical farmers who visit us every year at Malabar. It is, however, not written for the practical farmer alone but for those hundreds of city and suburban dwellers (many of whom come from farm or small-town backgrounds) who have a lively and even recreational interest in reading about everything which concerns agriculture, wild life, nutrition and in general that vast and complex world with which good agriculture is concerned.

The visitors to Malabar, bringing their own observations and knowledge, plus any number of stimulating queries and speculations, are one of the richest rewards of our life, our experimentation and our observation. Gradually Malabar Farm has become a modest sort of international trading post of agricultural knowledge, observation and of the speculation from which scientific discovery inevitably arises. The visitors come from all parts of the United States and Canada, and indeed from all over the world, so that it is possible within a single day to find farmers from Texas, Illinois and Ontario, exchanging experiences and observations with farmers and landowners from Brazil, India, Palestine and Europe. Indeed, before the Iron Curtain slammed shut, we had visitors from Soviet Russia.

Sometimes they come singly and sometimes as members of a group —a Farm Bureau Council, a Grange, a G.I. Vets' Class, a 4-H Club or a Future Farmers of America group—and frequently we have busloads of just good, plain, hardheaded and intelligent farmers who come from a great distance and spend a couple of days with us, stopping overnight at the nearest tourist camp. And there are county agents, soil conservation technicians and foresters, and the groups from the state agricultural colleges and universities, including the professors of agronomy, animal husbandry, and other specialized fields of agriculture.

At Malabar we hear from time to time echoes of such remarks as: "Of course they can do all that at Malabar because they have plenty of outside money."

Nothing could be less true, first because we do not have plenty of outside money, and second because, since the beginning of the enterprise, there has been in operation a strict rule that nothing is done at Malabar, either in the barns or in the fields, which any farmer could not afford to do. Indeed, there are many profitable practices which he cannot afford not to do. Even if the records did not exist on paper, they are there in the fields and the buildings for any farmer to see, and one does not deceive a smart farmer about costs. And there are the neighbors who, passing daily through the roads that intersect the land at Malabar, know how much fertilizer is used, how many fittings the soil has suffered and what the cattle are fed. The primary purpose of Malabar Farm has always been to establish facts and practices of use to the average farmer and even to the farmer who cannot get money from his bank but must go to government agencies for it.

The best evidence of these assertions comes from the practical farmers themselves who return home from a visit to Malabar to send back another group from the home county within a week or two, thus providing an apparently inexhaustible, and at times certainly an exhausting, stream of visitors. The most common compliment paid us is that we have provided a farm which is not made up of experimental plots or a place where one must go separately to a variety of different specialists to find the information desired, but that Malabar is a pattern of a farm with all the intricate difficulties of soil, water, weather, disease, economics, animal husbandry and many other things included in practical working order.

Together we tramp many miles a year across the fields and woods of Malabar, examining the soils at close range above and below ground. We observe the fashion in which the plants themselves warn you of the deficiencies of the soil in which they are growing. We spend many hours in studying the gravel pit where, in the glacial drift, much of it carried to our Ohio farm all the way from the Arctic Circle, the geology students find their special rich delights and where the farmers can observe what is going on beneath the surface and inside the earth itself. Here they can see the deep roots of the alfalfa, the sweet clover and the deep-rooted grasses, probing downward to a maximum depth of twenty feet or more, finding the rich minerals and converting them into organic availability, as the virgin Ohio forests used to do, and carrying them upward to the surface to nourish other plants and animals, and finally people.

The visitors come by the thousand, all welcome save the empty-minded, addlepated sightseer. Even among those I have observed converts who, finding themselves on the fringe of a group of farmers and scientists all talking together, have become interested and perhaps carry off with them a spark which will tie them perhaps in a small way into the vast complex web of the universe and enrich incredibly lives otherwise poor in the resources of living and hitherto occupied largely by the comics, the sport pages and the soap operas. We, too, at Malabar have learned enormously from the horny-handed intelligent dairyman and farmer with an imagination and an observing eye and from the professors and experts in research who explain the why of things or argue, rarely, that they know all the answers and that there is nothing more to be discovered or learned.

Often the visitors stay long after the big dairy herd has been milked and the cows are streaming down across the bluegrass bottom along the creek to disappear for the night among the misty willows. Often they stay until darkness falls, under the big walnut tree on the lawn of the Big House to discuss the most fascinating and complex and perhaps important of all professions which reaches out into every field of science from entomology and nutrition to economics and physics. Usually when the bull session at last breaks up, someone asks, "Isn't there some book in which we can get all of this in one lump? All of what we have been talking about today"—the observations of modest men and the speculation of the imaginative ones. Usually I reply that so far as I know there is no such book and there never will be because the subject of agriculture is so vast, so intricate and so complex. To write a comprehensive book on agriculture is like writing a book about the universe itself.

Certainly this book is not such a one. I have merely attempted to set down here some of the observations and speculations of the people at Malabar who make up a rare group, some with the hard economic approach, some with the approach of a John Burroughs or a Fabre. All of us love the fields and love our cattle. Most of us are simple people who love what we are doing, and that is perhaps the greatest satisfaction in life one can have. Also included in the following chapters are many of the observations and speculations of the visitors, from dirt farmer to scientist, who visit us each year.

I have tried to tie together in their proper relationships the many and varied facets of the New Agriculture, and in so intricate and complex a profession, I am aware that there are some repetitions and cross-writing, for in agriculture nothing is unrelated to anything else. Some of the passages in this book amplify those in *Pleasant Valley* and *Malabar Farm*, two books which were meant to lure readers who had never had any interest in agriculture or whose interest had been dulled or killed by pamphlets bearing such names as *Some Preliminary Observations of Phenological Data as a Tool in the Study of Photo-periodic and Thermal Requirements of Various Plant Material*. That is a good enough title for the pamphlet of one professor written at another professor, but the great problem of our time is to get information to the livestock man, the practical farmer and even to the peasant of Middle Europe and the coolie working his rice paddy on the terraces of Java and Burma. They are the ones who must feed us all and lower the prices of good food and diminish the taxes which, throughout the world, are devouring man's initiative and independence and human dignity.

And so this book is written in simple and direct language, as simply as possible, not only for the possible benefit of fellow farmers and livestock men but also for the great numbers of people who, without practicing professionally in either of these fields, still maintain a lively interest in them.

I hope, humbly, that the book will contribute something to the kind of good and solid agriculture which can stand squarely upon its own feet without the subsidies and the price floors and the maze of bureaucratic controls which serve, not to improve agriculture or in the long run even the economic condition of the farmer, but to subsidize a wretched agriculture and support the greedy speculator, farmer and broker alike, in overproduction or scarcity and continue the process of destroying the greatest continually renewable natural resource and the greatest base of *real wealth* which we, as a nation, possess.

One of the silliest and most superficial and cynical assertions ever made in the history of this country was the now happily dated assertion that "anybody can farm." Never, certainly, has there been a saying so completely devoid of truth in this country or elsewhere throughout the world. A good farmer has to know more about more things than any man in any profession now practiced. The belief that "anybody could farm" has cost us billions of dollars in taxes, in high prices and the destruction of the soil which is the fundamental and ultimate base of the wealth of every nation.

No really good farmer wants to be "kept" nor does he need be. One of the greatest satisfactions of the farmer's life is his ruggedness and his independence. When that is lost and he takes orders from the bureaucrat, the very core of his pride and satisfaction has rotted away and we shall arrive at an agriculture as poor as that of the collective farms of Soviet Russia where the farmer is no better than a slave who lives at a slave's level of food and shelter and where, in order to force the production of food, the state perpetually threatens the farmer with imprisonment, exile or worse.

In writing this book I have thought many times of Margaret Fuller's grandiloquent assertion, "I accept the Universe," and of Carlyle's quick response, "Gad, she'd better!" There are no short cuts, economic or medical or scientific, where the laws of the universe are involved. One works *with* Nature, whether in terms of soil or of human character, or one is destroyed. That, I think, is a law which it would be well for all of us—economists, politicians, farmers, Marxists, business men and all the others—to keep perpetually in mind. It would be well for man to contemplate daily the principal fact of his brief existence—the fact of his colossal physical insignificance.

L. B.

CHAPTER I

Out of the Sea

I am perfectly aware that it would be preferable to avoid repetitions and to give a complete story every time; but, in the domain of instinct, who can claim a harvest that leaves no grain for other gleaners? Sometimes the handful of corn left on the field is of more importance than the reaper's sheaves. If we had to wait until we knew every detail of the question studied, no one would venture to write the little that he knows. From time to time, a few truths are revealed, tiny pieces of the vast mosaic of things. Better to divulge the discovery, however humble it be. Others will come who, also gathering a few fragments, will assemble the whole into a picture ever growing larger but ever notched by the unknown.

And then the burden of years forbids me to entertain long hopes. Distrustful of the morrow, I write from day to day, as I make my observations. This method, one of necessity rather than choice, sometimes results in the reopening of old subjects, when new investigations throw light within and enable me to complete or it may be to modify the first text.

—EDWIN WAY TEALE,
The Insect World of J. Henri Fabre

I. *Out of the Sea*

LONG ago, so long that no even faintly accurate estimate of the time can be made by man, something happened in the vastness of the universe which has never been explained and which may never be understood. The event was of the utmost importance to all of us, for if it had not happened we should not be here at all breathing, working, reproducing our kind. In the vastness of cosmic time and space there appeared a tiny spark which was Life. It appeared first in the steaming waters of a planet which was in the process of cooling off. It did not appear until the temperature of the earth fell below the temperature at which it could come into existence and survive. It appeared first in the vast waters which covered most of the planet, and very likely the spark was contained in a single cell which presently divided into two cells containing two sparks of life, and so on as it multiplied, or in the beginning there may not have been even a cell but only a kind of substance with an impetus to live.

As time went on—time in terms of millions of years—the little one-celled organism or the inspired substance turned into a more complicated structure with the spark of life contained in three or four cells and then millions of years later into organisms containing many cells which ceased any longer to be simply soft amorphous shapeless little objects attached to rocks or sand or able only to move aimlessly by a certain wriggling movement, and the multiple cells began to take shape, some of them in the form of living things which remained rooted to the cooling rock or sand and developed a process of survival, largely upon sunlight, air and water, drawing from the slowly crumbling rock the minerals which made this process possible. Others moved about and began, again through millions of years, to take various shapes better adapted than the aimless wriggling to propelling themselves about in search of food. They in turn developed an immensely complicated process by which they turned into nourishment and growth, not merely sunlight, air and water, but the living or dead substance of other organisms, both plants and animals. The first group became the plants and trees and the other vegetation of our existing world, and the second became the fish, the mammals and the birds.

The process by which plants live largely on sunlight, air and water with the aid of minerals and convert these into chlorophyll (the green part of plants) and into proteins and carbohydrates, which in turn sustain the life of man and his animal companions, is known to modern science as photosynthesis. The process has been named but we do not yet understand fully the mysteries of its operation.

The process by which man and his animal friends are able to turn plants and the meat and milk and eggs of his companions on earth into nourishment, growth and the means of reproduction is known as metabolism. The two processes are not far apart in their operation, probably much less divergent than we know today. Both still depend upon the minerals from the decaying primitive rock for the mineral substance which makes continuing life possible.

How that original spark of life came into existence we do not know and may never find out. The only suggested clue lies in the viruses of many diseases, non-filterable substances which are perhaps the very lowest form of life, preceding even that of bacteria and the single-celled organisms. But there are many things we do not know and understand about this world and universe, undoubtedly more things that we do not know than we know. Today, with the aid of the spectroscope, the electronic microscope, radioactive isotopes and other instruments born of the ingenuity of man, we are daily making more discoveries than the mind of any one man can record or analyze or co-ordinate. Many of the puzzles could perhaps be solved by the co-ordination of these multiple and complex discoveries. But man is not good at co-ordination. He is too much the individualist, and he has made himself frequently into the specialist who goes down a single blind alley, believing that he alone has found all the answers to the universe within his own narrow field. Sometimes he is the evangelist who, fired by a single idea or discovery, goes off preaching scientific salvation by the flinging about of unproven assertions which he heralds as unassailable facts. And sometimes he is the academic professor sitting in his study armchair surrounded by dusty and obsolete scientific books, pompously asserting that all has been learned and that there is nothing further to learn in a universe of such complexity and variety that it is quite beyond the comprehension of man and concerning which we have until now learned only a tiny fraction of what there is to know.

Within every true artist or farmer or scientist there is a spark, as

[4]

precious as that first tiny spark with which life itself began, that is compounded of imagination and curiosity and speculation, which are the handmaidens of creation. There is as well the immensely important faculty of observation, that kind of observation which revealed to Fabre in his tiny backyard in Provence a picture of the universe itself. The closed and academic mind is essentially the undertaker of science and knowledge and winds their shroud. In the long history of mankind, the tiniest observation or speculation of the most humble men (and all really great men are humble in the face of Nature) has sometimes led to vast and dynamic discoveries of the utmost importance to man. Many of the greatest contributions to agriculture in our time have not come from the billion-dollar Department of Agriculture nor from the countless colleges of agriculture but from a county agent or a farmer who had the power to observe, the imagination to speculate and the logic to deduce a process from which vast benefits have developed.

I have quoted in earlier books the saying of Confucius that "The best fertilizer of any farm is the footsteps of the owner." Put another way, the best farmer is the man who knows every foot of his farm, the condition of the plants and of the soil and has the capacity to learn as much from his plants, his soil and his animals as any college of agriculture can teach him; and what he learns from these elements will not have to be *unlearned* later on because some evangelist jumped at a conclusion or because some specialist asserted as a universal law some truth which had a local or regional basis but failed as truth under other conditions of soil or climate or diet in a world in which the variety of all these things is infinite.

The good farmer is the man who learns as much as he can about the vast range of things which the good farmer must know concerning veterinary science, economics, chemistry, botany, animal husbandry, nutrition and countless other fields, all of which are tied into the ancient, complex and varied profession of agriculture, and then knows how to apply this knowledge to his own problems.

Perhaps no stupider human saying has ever been formulated than the one that "anybody can farm." Anyone can go through the motions, but not 10 per cent of our agricultural population today could be seriously called "good farmers." Thirty per cent are pretty good, and the remaining 60 per cent do not, through ignorance or laziness or sometimes through the misfortune of living on wretched land fit only for forests, deserve the dignified title of "farmer." Most of them still remain within the range of a completely primitive

[5]

agriculture confined to plowing, scattering seed and harvesting whatever crops with luck turn up at the end of the season. That they perform these operations with the aid of modern machinery does not make them either good or modern farmers. Tragically, a great many of them actually hate the **soil** which they work, the very soil which, if tended properly, could make them prosperous and proud and dignified and happy men.

You might ask, how is all this concerned with that infinitely tiny spark of life ignited millions, perhaps billions, of years ago in the still steaming waters of this planet? The connection, as we are discovering more and more thoroughly each day, is a closer one than appears on the surface. Out of that tiny spark and the first cell or substance which contained it, all of us were eventually descended, through the myriads of strange monsters which embodied and carried on that tiny spark as cellular structure multiplied and life began to take strange shapes, each with its special adaptations for life and survival and reproduction in a given environment.

As cells multiplied, they became fish, and fish crept ashore and developed lungs and became lizards, and finally birds and horses and camels and elephants and man in his way followed a similar transmogrification, the long record of which appears in the embryo all the way from the gills of the fish to the lungs and the formation of mouth and throat which give him the power of speech and the brain which has raised him gradually above the level of the other animals. The whole record is there—a record which none of us can escape, however much we may try to deceive ourselves and avoid our past.

In the beginning we all came from the sea, and the elements of the sea—its oxygen, its nitrogen, its infinite range and variety of mineral wealth—are still as necessary to us as the proteins and carbohydrates which they make possible and by which we sustain life. Some of them we are able to live without as we are able to live without gills. Some of the organs and glands once necessary to existence and procreation have become atrophied and useless and will doubtless disappear one day as our gills have disappeared, but we are, like the plants themselves, still dependent upon a remarkable range of minerals for the structure of our bodies, for the vigor and health and the maintenance of that metabolic system which permits us to utilize the complex proteins and carbohydrates which give us energy and brains, and for the fertility which permits us to reproduce ourselves and so continue the future of a species which is still in the process of change and adaptivity.

Once not very long ago it was believed that only phosphorus and calcium and one or two other major elements were necessary to the development, birth and growth of a living human organism. Today we are becoming aware that in the functioning of a normal and healthy metabolism producing a healthy, vigorous human, or even an animal or a plant of the same category, an infinitely greater range of elements and minerals is necessary; and constantly, almost day by day as new discoveries are being made, the range is growing larger to include more and more of the minerals from the sea out of which that first spark of life was born.

The relation of iodine to goiter and cretinism is perhaps the oldest and longest-established case. A child born in a wholly iodine-deficient area stands one chance in ten of being an idiot, one in five of dying of goiter and virtually one in one of suffering from all the maladies and handicaps that come of the disorder of the thyroid gland, all because he and his mother before him did not have daily as much iodine as could be smeared on the head of a pin. The function of fluorine in the most minute quantities in the creation and maintenance of good teeth and bone has long been established, or that of cobalt, copper and manganese in relation to acute anemia and the capacity to breed. Cobalt in exquisitely minute quantities plays its part in the creation of one of the most recently discovered and important of all vitamins, B-12, a vitamin by which it indeed could be said that we live.

But as the physiologists and men of medicine are beginning to discover, the story of life, complicated as it is, contains still greater complexities, many of them still undiscovered. One may suffer from thyroid derangement through the lack of infinitesimal amounts of iodine, but the thyroid derangement may make it impossible for the human metabolism to absorb sufficient amounts of calcium or phosphorus, no matter the amount taken into the body, or impossible to absorb the infinitesimal amounts of zinc which may be the safeguard against leukemia. The pattern is intricate and extremely difficult to unravel—this pattern of nutrition and the relation of minerals to our health, growth, intelligence and vitality, and to the intricate workings and interdependence of all our glands. We have only begun to unravel the fringes of the whole pattern of metabolism, of minerals and vitamins and enzymes and hormones and the nutrition which, it appears, runs back and back into the steam and fogs of the primeval world in which the first tiny spark, born of the minerals and elements themselves, came into existence.

Why, you might well ask again, should this concern the farmer? It concerns the farmer because the chief concern of the farmer is his soil. Out of it comes the health, intelligence and vitality of his animals and his family and, in a broader sense, of his fellow citizens. Out of it also comes his economic prosperity, his independence of banks and of government subsidies and regimentation. If his soil is good and minerally balanced and well managed and productive up to the optimum (which means simply that he is getting maximum potential production in quantity and quality without loss of fertility) he is the most independent man in the world, a world which has never been able to do without him and which becomes daily and hourly and by the minute less and less able to do without him. Slaves do not produce great quantities of food (as Soviet Russia has discovered) nor do they produce, except by accident of Nature, good and highly nutritious foods. The good independent successful farmer produces both. Better than any man he knows, through his plants and animals and his daily contact and struggles with the weather, that out of the earth we come and to the earth we return. Out of the sea, one might almost say.

The men and women of no other profession are as content to die when their time comes as the good farmer and his wife, for, better than the people of any other profession, they know by living with earth and sky and in companionship with their fellow animals that we are all only infinitesimal fragments of a vast universe in which the cycle of birth, growth, death, decay and rebirth is the law which has permitted us to live.

What we have done with our individual lives is another question for which we ourselves must accept the responsibility. If it has been a good life, full and rich, and especially if it has been lived close to the earth from which we come, there are no terrors and no yearnings for a silly heaven of pink clouds filled with angels twanging harps. Rarely does the good farmer long for any immortality better than the rich fields he has left behind him and the healthy, intelligent children who will carry on his work and his name. If there is an after-life that is pleasant and comfortable, so much the better, but he hopes that it will not be an after-life in which there is no work, for it is by work of hand and brain that he has lived a full, rich life, which leaves him at the end ready to lie down and fall asleep in the quiet knowledge and satisfaction that what he has done will live on and on after him into eternity. The good farmer is one of the ultimate peaks of evolution away from that first silly one-celled creature wriggling about in the sea water of a billion years ago. He is the one citizen without whom mankind and civilization cannot exist. As that very great and vigorous old gentleman, Liberty Hyde Bailey, has put it so well, "The first man was a farmer and the last man will be a farmer."

CHAPTER II

The New World in Agriculture

A *cubic foot of soil must be a big subject. During recent years I have asked three different, highly learned experts, men whose names you have heard, to write an article. Just take one cubic foot of sod and tell* Farm Journal *readers all about the insect and microbe life, about the geology and chemistry of the soil particules and everything else there is to know. None has been willing to undertake the assignment. Maybe I had better try to get someone to write about just one cubic inch.*

—Wheeler McMillen,
Editor, The Farm Journal

Dear Wheeler: I think the trouble is that as yet we still do not know what a cubic foot of productive living soil really is. The other obstacle lies in the overspecialization of our education. The chemist could probably tell to you his side of the story and the entomologist his side, the nutritionist his side and so on, but there seems to be no one alive today who can put it all together and give us the answer.

Always yours,
Louis Bromfield
P.S. Reducing the dimensions to one square inch wouldn't make much difference. It would still contain billions of living organisms—if it is truly productive soil.

II. *The New World in Agriculture*

MOST scientists concerned with agriculture and all good farmers would, I think, agree with the statement that more progress has been made in agriculture and more knowledge acquired concerning soil during the past generation or less than in all the history of the world up to now. I also think it conservative to say that we probably know only about 5 per cent of what there is to know concerning soil and its relationship to ecology, to economics, to nutrition, health and intelligence, to sterility and countless other factors closely related to the well-being of the creature known as man.

Some of this knowledge is merely confirmation of age-old agricultural, nutritional, and medical principles and practices, many of which until quite recently have been regarded as superstitions. Other knowledge and information has resulted from experiment and research which has demonstrated, beyond any dispute, that good productive soil is something immensely complex, with ramifications touching the health and very existence of plants, animals and men, which we are only now beginning to understand.

There was a time in the history of this country when agriculture consisted very largely in the simple processes of plowing, fitting, seeding and harvesting. This was, of course, a very primitive agriculture and one which could succeed for any length of time only upon the better virgin soils provided by Nature. It belonged to a period which is long past, but the fact that there are still many farmers and absentee landlords and tenants, particularly in our backward areas, who, through poverty or ignorance or greed, persist in this form of agriculture, costs the American people hundreds of millions annually in farm subsidies, in parity supports, in government buying to maintain prices and in floor price guarantees, as well as high prices across the counter.

The first error in American agriculture arose from the silly idea that "anybody can farm." With the knowledge we have today, we know that nothing could be less true and that the theory has cost the American people billions of dollars.

Following this school of thought came another heresy which like-

[13]

wise worked great damage not only to the soil and to our natural resources but to the individual farmer in the nation as well. It was the assertion and even the belief, spread far and wide by commercial fertilizer salesmen and many contemporary soil "experts," that in order to produce bumper crops, achieve prosperity and health and maintain fertility of the soil, the only necessity was a mixture of potash, nitrogen and phosphorus, held together in various chemical combinations. The single concession made by the salesmen and some college professors was that a little ground limestone or calcium in other forms might be helpful. During the period when these beliefs and teachings held sway, the structure and productivity of many soils over great areas deteriorated steadily and constantly, until the soils became little more than a cement of unavailable minerals with traces of acid, highly susceptible to erosion and destructive to proper drainage.

This school of thought ignored completely the whole factor of organic materials and of all the fungi, moulds, moisture, bacteria, anti-biotics, vitamins, enzymes, hormones, earthworms and many other things which are essentially a part of any living and productive soil and which maintain and increase their fertility and their life and health-giving properties. On many soils the results of this ignorance were utterly disastrous, for this easy "short-cut" method of agriculture, in the long run little better than the "anybody can farm" school of thought, created devastating erosion, ruined drainage and constantly reduced the availability not only of the natural fertility of the soils but actually the availability of the commercial fertilizer itself. One found the results at their worst in the corn and cotton single-crop areas where farms presently arrived, under the practice of such an oversimplified agriculture, at a point where a farmer might spend $100 on fertilizer and not derive $10 worth of benefit from it because his soils had lost their texture and in the hot growing season were unable to absorb or maintain enough moisture to make the commercial fertilizer available.

Countless farmers accused the fertilizer of "burning out" their crops or spread the report that commercial fertilizers were not worth the money paid for them, neither of which statements was true. The fertilizer was useless or damaging only in inverse ratio to the amount of organic materials and humus in the soils. A cement-like soil could neither absorb rainfall nor conserve and maintain either moisture or drainage. The disastrous results of this "short-cut" school of thought presently brought on the four- or five-year rotation of oats,

[14]

corn, wheat, grass and legumes in order to restore through organic material the texture of the soil and its capacity for absorbing and maintaining moisture. But even the four- to five-year rotation was uncertain of results on light soils and in hot climates and in many cases has proven inadequate for increasing or maintaining moisture, organic material and true fertility. And many soils have reached so ruinous a condition that it requires the most intensive use of green and barnyard manures and sods to restore them even to a moderate level of production.

The whole commercial fertilizer theory represented both the ignorance and the arrogance of the limited or greedy men, manufacturers and farmers and professors, who are perpetually seeking a short cut or a means of outwitting Nature and the very laws of physics, of chemistry and even of economics. In agriculture or animal husbandry or horticulture or in any profession based upon the laws of the universe there are no short cuts. Working within the laws of Nature and with her, certain of her processes, such as the production of topsoil or the availability of mineral fertility, can be immensely speeded up, hundreds perhaps thousands of times, but such an operation is totally different from the "short-cut" school of agriculture which is closely related to the old Snake-Oil seller who had a preparation good for any disease of man, child or beast, which would cure everything from the seven-year itch to cancer.

Man can neither outwit nor short-cut natural law, but by working with it he has achieved wonders ranging all the way from the atomic bomb and television to 200-bushels-to-the-acre corn and cows which give 100 pounds and more of milk per day. No man ever "licked" Nature and natural law, but countless men, especially farmers, have ruined themselves physically and financially by fighting her and trying short cuts and cure-alls.

We know today a little concerning what a cubic foot of good, living, productive, profitable soil really is and we even know how to produce it. We could produce good productive soil out of a brick or a cement pavement or blue shale or granite or sand. The only limiting factor is the economic one—that it would cost much too much; and so we must work with soils which, although sometimes depleted, leached or unbalanced, may still be made living, productive and profitable at a reasonable initial cost plus investment.

We know today that a cubic foot of good, productive and highly profitable soil is actually a living thing in which all the laws of the universe are constantly in operation. It contains not only a conglomeration of the minerals necessary to life and out of which life itself was born and is still maintained, but it contains myriads of living organisms from minute bacteria to the earthworm which is a giant by comparison. It contains as well both growing and decaying organic materials together with certain organisms and strange growths about which we know as yet very little, although each day brings astonishing and miraculous new discoveries concerning their character, properties and functions—such things as the whole range of acids, enzymes, hormones, anti-biotics and vitamins existing in good living soils or actually born of them. Beyond these elements there are the glandular secretions, hormones, bacteria and enzymes passing into the soil through the employment of barnyard manures and having a very considerable effect upon the germination of seeds, the creation of proteins and carbohydrates and very possibly, through creating a much higher availability of minerals, an effect upon the very processes of photosynthesis in plants and metabolism in animals and people.

Ever present in a cubic foot of living, productive soil is the cycle which is a fundamental law of life on this planet and perhaps eventually of the universe—the cycle which includes birth, growth, death, decay and rebirth. In short, there exists in a cubic foot of good soil the vestigial chain by which man developed through billions of years out of that faint spark of life which appeared spontaneously in the tepid waters of a rapidly cooling earth. By the time man has consumed for his own nutrition and existence the vegetable products of that soil and the dairy products, milk, eggs and meat, in turn created out of the same vegetable growth, the whole chain becomes complete.

It is a chain involving so many elements and factors and the interplay of so many operations in an intricate pattern as to render even the imagination confused and dizzy. It involves the anti-biotics, such as penicillin, streptomycin, aureomycin, chloromycetin, all born of moulds and fungi and most of them produced only in rich soils with high organic content and preferably containing animal manures. It involves the whole field of enzymes and hormones and vitamins and the effects, almost incredible and certainly miraculous, of certain trace elements upon definite glands, and the capacity of rich and productive soils and benevolent bacteria to destroy malignant disease germs and to build a nutrition for plants, animals and people resistant to disease and even to the disintegrating process known as old age and to the degenerative diseases arising from that process. It is not impossible that when the secret of life and its beginnings is discovered, the secret will be found in a cubic foot of good, rich, productive soil. We have, in agriculture and in soils, barely lifted the edge of the curtain.

In the face of all these potentialities and actual discoveries it is manifestly ludicrous for any man to claim that we have learned all there is to know about agriculture or that "anybody can farm" or that applying quantities of phosphorus, nitrogen and potash in chemical form mixed with a cement-like "filler" is the whole and only answer to a sound and profitable agriculture.

At Malabar Farm we have been led by our own soil, by the results we have obtained and by the factors we have observed, virtually to assume that the laws which govern good soil and optimum production are perhaps as exact and immutable as the laws of chemistry, physics or any of the natural sciences which play so important a part in the creation of such soils. We are inclined to believe that there are a series of balances, absolute in character, which, when attained, produce optimum production, which is simply maximum production both in quantity and nutritional quality without reducing and possibly even augmenting the fertility factor.

We are inclined to believe that production of soils, both in terms of quantity and quality, mounts or declines in exact ratio to the degree with which their absolute balances are established and maintained. Any intelligent farmer is well aware that when his major elements—calcium, phosphorus, potash and nitrogen—are out of proper balance for any given crop, lowered yields and poor quality results. Sometimes the unbalance leads to actual deformation of the plants and to all manner of signs, as clear as the spots of measles on a child, of deficiencies, or in the case of excess nitrogen to a growth which is rank enough but inferior in quality and in seed and reproductive capacity.

Total lack of any of the major elements leads to the production of a diseased and crippled plant life and eventually to diseased, crippled and infertile animals and people. The total lack of even such a minor element as boron can cripple plants and fruit trees to a point where they become virtually sterile and eventually wither away.

In our own observation at Malabar, there seem to exist three major balances of first importance: (1) the balance between minerals and organic material; (2) the balance among the four major elements—calcium, nitrogen, phosphorus and potash; (3) the balance between these two major balances and a whole balance of so-called trace elements including manganese, magnesium, boron, iodine, fluorine, bromine, copper, sulphur, cobalt, molybdenum and many others, possibly including the whole range of known elements which were originally present in the sea water out of which life was born. These are called trace elements because in order to produce optimum crops and consequently optimum production of animals and people, both in quantity and in quality, health and vigor, they need to be present in the soil only in small quantities, some of them to an almost infinitesimal degree.

Research has established the fact that when iodine, even in minute quantities, is not present, disease and malformation result; when cobalt and copper and manganese are absent, animals and people become anemic and sterile; and when fluorine is present in water at the rate of one part per million (provided there is also abundant calcium in balance) tooth decay in man becomes virtually unknown at any age.

These balances we have found to be of the utmost importance in the production of crops, animals and people both in quantity and quality. Most good soils contained most of them originally, and in many soils, especially those which have been well farmed with abundant animal manure and with some effort made to replace the drain in the thin top layer of the minerals carried off in the form of grains, milk, eggs, bone and meat, they are still present.

To the layman without experience in agriculture or soils, the phrase "virgin soil" implies a rich, well-balanced and productive soil. Few implications could be less true, for Nature laid down her soils in a haphazard fashion without regard to the balances in question. Very few soils contain the major elements, the organic material and the trace elements in any degree approaching the balance which results in optimum production. Perhaps only in the sea itself does every element exist because all elements, through erosion and run-off water, eventually reach the sea from which life itself came. Yet even in the sea there are serious unbalances in relation to given vegetations. Salt marshes are infertile save for a narrow range of specially adapted and qualified plants, largely because their soils and the waters which feed them contain too much chlorine and too much sodium—a notable case of exaggerated balance. Indeed, sea water cannot be used for irrigation purposes because of the notable unbalances which actually render it destructive and even toxic to most plant life. Yet we know today how, by complicated chemical processes, to rearrange the balances of minerals existing in sea water to a point where it may actually behave as fertilizer.[1]

[1] This is actually the process behind the production of the Sea Soil turned out by the Dow Chemical Company at its Freeport, Texas, plant. The Sea Soil

Specifically the expression "virgin soil" means simply soil which has never been put under cultivation by the hand of man, and some of our "virgin soils" are among the worst in the world through deficiencies and *unbalances* for which Nature herself was responsible in the process of laying them down. Areas of Wisconsin and Michigan and Minnesota were deforested or drained with the idea of putting them to agricultural production, only to be discovered to be of such poor balance, or so depleted by leaching caused by their sandy quality, that they could not raise crops decent enough to pay taxes or interest, and the cost of artificially creating out of them soils which could be even moderately productive was economically prohibitive. Those areas have been rapidly reforested or restored to a swampy condition as refuges for wild life or have been allowed, uneconomically, to grow back as second-growth, low-quality forest.[2]

There are areas in Florida and along the whole of the Gulf Coast

---

is largely the residue left from sea water after the excess amounts of chlorine and sodium and magnesium have been removed for chemical and industrial purposes. The residue contains every mineral, even to traces of gold and silver, that could be expected, but few of them in an unbalance or a quantity which might prove toxic to a wide, given range of plants. On the contrary, we have every evidence at Malabar that the application to the soil of this residue of elements, in which there still remain traces of both sodium and chlorine, is highly beneficial. We have rose beds where the Sea Soil was applied in which the plants are not only notable for their vigor and the deep green of the foliage and the perfection of blooms but were *totally* immune both to disease and to insects without dusts or sprays during the whole summer of 1949. It should be remembered also that plants exert a definite selectivity and do not of course use up *all* the available minerals and elements in a given season but only the amount required for their well-being. It is only when there is an exaggerated amount of given mineral present or a total deficiency that plants display the toxic or crippling effects of either condition. Animals, of course, also show a similar instinctive and highly selective capacity when given access to a variety of essential minerals. This is especially so when the forage on which they are feeding is deficient and they are given access to minerals in the form of chemical salts.

[2] During the Great Depression the Great Swamps of Minnesota were drained by W.P.A. workers at vast expense, only to show that the soil was wholly unsuitable for agriculture. By draining them, a vast area in which every kind of wild life had flourished was destroyed. The only solution seemed to be to restore the swamps to their original condition, but this project encountered difficulty in an already existing law which forbade the blocking of any drainage ditch by an artificial barrier. A solution was finally discovered through the introduction of beavers which promptly went to work damming up the ditches and reproducing more beaver engineers at a tremendous rate. The restoration of the swamp wildlife area has practically been accomplished, less expensively by many millions of dollars than their original draining had been accomplished by the W.P.A. This kind of ecological ignorance and stupidity—the draining or cutting over of land useless for agriculture thus destroying water tables, springs and wild life—is common enough in the United States. It may be observed in almost any state.

where one can see cattle walking about all day in rich-looking grass up to their knees with their ribs and hip bones sticking out. These wretched animals are a notable example of the results of unbalances in soils and of the deceptiveness of a lush growth which appears to the eye of the untutored to be nutritive but is not because of deficiencies of almost everything but nitrogen, carbon and water.

This grass, despite its lush quantity at certain seasons, is not an example of *optimum* production. There is quantity, but on the side of quality the whole principle of optimum production breaks down. The cattle simply cannot eat, contain and digest enough of the abundant but deficient vegetation to supply their nutritional needs and keep them in good flesh. Moreover, there is also no doubt that the deficiencies also affect the functioning of their glands and their metabolism, creating in turn, within their organs and glands, an inability to absorb and utilize the scant, unbalanced mineral content of the forage which they consume so ravenously. They are the perfect counterpart of many people living off the produce of deficient and unbalanced soils, who are starving to death while gorging themselves.[3]

An exact contrast and the converse of the Gulf Plain conditions exists only a little way off in the almost arid regions of the Great Plains where cattle, wandering over land sparsely covered by tufty grasses which appear, both in winter or summer, far from lush or green, remain fat and sleek and healthy because the Great Plains as a whole is an area notable for the fine mineral balance of its soils and, despite a rainfall which is only a fraction of that on the damp Gulf

[3] William Albrecht, Professor of Soils at Missouri University College of Agriculture and one of the greatest living authorities upon soil and nutrition, has urged for many years that the measure of yield on an acre of land should not be that of bulk production but of the pounds of meat or milk produced by the feed grown upon that acre. We should thus have a measure based not merely upon *bulk*, which, as in the case of the Gulf Plains land, would be wholly misleading in the sense of real nutritional and productive values, but upon a blend of quantity and quality which approached the optimum. Hybrid seed corn, which produces through the use of hybrid seed alone an average increase in yield of 20 to 25 per cent, has also shown a steadily declining content of invaluable proteins. This is a case of gaining bulk or quantity while sacrificing quality. In this case it is highly likely that the declining protein values do not arise from the *hybrid* element but from the fact that in most of our corn land the soil structure and consequent availability of minerals which create the protein factor, have been steadily declining and are still doing so. In some cases, of course, under a poor agriculture, the soils have actually become depleted of the vital minerals themselves. Here again is a case in which the disruption of balances, notably in the range of organic material in the case of corn farmers, can bring about not only a decrease in bulk production but in nutritive quality as well.

Coast Plain, produces in its grasses and legumes a much higher level of mineral and vitamin nutrition. In bulk, an animal need consume on the Great Plains only a fraction of the weight of the Gulf Plain forage in order to remain sleek, healthy and in good flesh.

It is also highly probable that, out of the well-balanced and highly mineralized soils of the Great Plains (which have never been leached out by the heavy persistent rains), the animals are getting those minerals which activate the physiological functioning of their glands and make it possible for them to assimilate, under a perfectly working metabolism, the maximum nutritive value of the forage they are consuming.[4]

In this case, as in so many others related to the question of soils, organic material, balances and nutrition and their effect upon the functions of glands and metabolism, the processes are not simple but infinitely complicated, with innumerable factors working upon each other to create new factors in the complex process which creates and maintains life on a sound and high level, whether in plants, animals or people. Many of these complicated processes are today wholly

[4] In this problem of the relation between nutritious mineral balance and mere bulk of forage or feeds, there enters another important element—that of availability. In some portions of the Gulf Coast, which is throughout notable for its deficiency of phosphorus, there is an abundance of other minerals actually in the soil, but for some reason these are unavailable and do not show up in organic form in the vegetation and therefore are, of course, unavailable to the animals feeding on it. This is one of the countless mysteries, as yet unsolved, which lead to the conclusion that, with all our great advance in knowledge of soils, we still know perhaps not more than 5 per cent of what there is to be learned. In some Gulf Plain areas there is an actual superabundance of calcium to the point of being out of balance even for the demands of legumes, yet it for some reason is largely unavailable to the vegetation. I have noted, however, that wherever oyster shells have been applied, usually accidentally, to such soils, the growth of legumes is greatly stimulated, and without the usual signs of deficiencies, indicating that by some process the lime has been made more available. This is true even though, in adding oyster shell, one is adding further calcium to a soil in which the calcium is already overabundant and actually out of balance but not available. Either the increased vigorous growth of the legumes arises from the highly available calcium contained in considerable amounts in the oyster shell itself or there is present in the oyster shell other minerals, notably magnesium, which tend to make the locked-up lime available. Good farmers in the Middle West have told me many times that application of magnesium to their soils appeared to increase the availability and therefore the value of the calcium already in their soils, either naturally or put there by application. Indeed, there are many evidences that magnesium is perhaps not properly classified as a trace element as has sometimes been done but should be included with the other major elements, nitrogen, phosphorus, potassium and calcium. It is also possible and indeed probable that in many areas of the Gulf Plain the unavailability of minerals is also connected with poor drainage and a high water table.

unknown, but in their workings we may eventually find the clue to many of the most tragic ills of the human race from the physiological causes of imbecility to cancer and the whole range of degenerative diseases. The reader will no doubt grow weary of the expression "we still do not know," for it is one that must be repeated again and again, since it is so completely true.

In Florida there are great areas which were originally badly unbalanced minerally or were wholly deficient in some minerals. When the Everglades region was first drained and put to growing vegetables in the winter season and cattle grazing in the summer, expectations of making great fortunes out of the deep black, drained muck were in the beginning grossly disappointed. The vegetables failed to mature properly and the cattle very nearly starved to death despite a rank-growing but almost nutritionless vegetation created out of water, nitrogen and carbon. It was not until virtually the whole category of minerals known as far as now to be necessary for optimum production had been put into the drained soils that healthy and profitable vegetable crops could be grown and cattle maintained as a profitable operation. As the county agent of one of the Everglades counties explained it while we were making a tour of the area, "What you are going to see here is not agriculture but only a glorified form of hydroponics [the science of growing vegetables in solutions saturated with minerals]. We have to put into the soils regularly an immense variety of minerals and elements in order to produce successful and profitable crops whether of vegetation or animals."

In the sandier areas of Florida where citrus fruit is grown, similar deficiencies brought on an infinite list of troubles in the forms of disease, sterility and even insect attack, and it was not until the necessary minerals, both the major and the trace elements, were put into the soils that the troubles abated and it was possible to grow profitable and nutritious crops of fruit with considerably reduced spraying costs. It is notable that the trace elements, especially manganese, boron, copper and cobalt, used in very minute amounts, played a great part in correcting the balances and producing something approaching potential quantity and quality production. It is also notable that immediate results in correcting the diseases caused by soil deficiencies of such minerals can be achieved not only by adding the minerals to the soils but even by spraying them in solution directly onto the plants and trees, which appear to possess the power of absorbing them.

For certain crops, the element boron, in very small quantities, is virtually a necessity. In the case of apples, when there is a deficiency or total lack of boron, it is useless attempting to grow fruit for the apples remain stunted and their texture becomes flavorless and acquires the consistency of cork. Yet as little as 10 pounds per acre of ordinary borax worked into the soil will serve to correct the sickness of the apples and 20 pounds will assure excellent fruit provided the other elements and minerals are in reasonable balance. Boron is also a necessary element, even in small quantities, in the production of good alfalfa, as the farmers and government experts have discovered over very large areas of the Deep South where, until quite recently, it was found difficult or impossible to get a seeding of good healthy alfalfa even after the hunger of this green-gold crop for phosphorus, calcium and potash had been satisfied. The application of a few pounds of ordinary borax per acre corrected the unbalance and alfalfa almost immediately gave big yields of healthy green forage, adding enormously to the permanent improvement of the soil and to the immediate economic income of the farmers in the area.

It should not be overlooked that minerals may get out of balance on the overabundant as well as on the deficient side. Of this boron, a trace element, is a notable example. Used in too large quantities it becomes actually toxic even to boron-loving plants such as alfalfa or the apple tree. It should be remembered as well that plants show a great variation of tolerance toward certain minerals.

Plants of the legume family have a tolerance and even a liking for considerable amounts of calcium, but the presence in the soil of even moderate amounts of the same calcium will kill such sour-soil plants as the rhododendron, the laurel and the azalea and actually poison such poor, sour-land vegetation as that represented by sorrel and broom sedge. It is well to consider what crop you wish to grow and what are the proper balances for its whole nutrition, since plants vary considerably in their special requirements. There are none, however, which do not suffer when there is a total lack of any of the elements now known to be necessary to good, balanced, productive soils.

It is, of course, obvious that acid-loving plants such as the rhododendron or the humble broom sedge and sorrel did not appear on the earth full-blown with a definite taste for acid soils. They were first of all developed out of acid soils and over centuries adapted themselves to these soils just as conversely the legumes developed upon soils high in calcium content and would and do die when transferred to highly acid soils. Any potato farmer knows that potatoes and alfalfa will not flourish equally upon soils with a fixed arbitrary balance, and he uses rye for his green manures rather than the legumes since the legumes would flourish best upon soils too "sweet" to grow first-class potatoes.

Like everything in agriculture, none of this is simple, and these various balances in relation to given crops are one of the many elements which tend to make obsolete the old-fashioned, general, four- to five-year rotation farm. A farmer may keep his soils in proper balance to grow potatoes or to grow legumes but he cannot do both, and in consequence his yields, the quality of his products and the amount of his income will suffer in comparison with the farmer who specializes and does a first-rate job with one or the other.

In our own truck gardens, where both experiment and observation are intensively practiced as a kind of a laboratory for the rest of the many acres at Malabar, the effect of these balances in relation to given vegetables is extremely noticeable, even to a point where we maintain two gardens and even parts of gardens in which the balances are adjusted to specific given crops, a practice followed, of course, by all successful truck and hothouse growers.

Nitrogen, of course, is the prime and obvious example. In one garden the nitrogen content for such leafy crops as lettuce, cabbage and celery has been raised so high that it is virtually impossible to produce good root crops such as carrots, beets, parsnips and radishes. Under the heavy nitrogen, these root crops go largely to leafage with very poor production below ground. It is also true that the excess nitrogen appears to upset the resistance of the root crop plants both to disease and to insects. On the high-nitrogen soils, the root crops are much more subject to attack than on the soils lower in nitrogen, apparently because the balances demanded by leafy crops and root crops vary so widely that, when the elements and minerals are out of balance in soils in relation to one crop or the other, their whole photosynthetic process is upset, they are weakened and fall victim to attack by both disease and insects. The soil in all our truck gardens is too high in calcium to produce first-quality potatoes, and it is notable that the potatoes suffer most from disease and insect attack of all the wide range of fruits and vegetables grown there.

These facts are, of course, well known to all first-class market and hothouse vegetable growers who are far in advance of our general agriculture in knowledge and practice in the field of organic material, trace elements and mineral balances. This is so of necessity because deficiency of even minute quantities of one of the trace elements or a serious imbalance of the major elements may cost such agricultural operators thousands of dollars in a season by producing stunted or defective vegetables unsalable in a market where there is nearly always abundance and sharp competition. A shortage of manganese, even in minute quantities, may, for example, render a whole season's crop unmarketable because of the blemishes, splitting, uneven ripening and discoloration which are the symptoms of deficiencies as much as goiter in man and hairlessness in livestock are symptoms of a deficiency of iodine in minute quantities.

We at Malabar have become more and more engrossed in the relationship of deficiencies in minor elements and, indeed, in all imbalances of elements in relation to the health of plants and their resistance both to disease and insect attack. By "sorting out" crops into different gardens and parts of gardens where the soil balances varied and a specific soil composition suited a specific plant, we have arrived at a high degree of resistance to both disease and insects, so that in one garden where only leafy vegetables are grown there has been no need to dust or spray against either disease or insects for a period of four years. There has been zero disease and no insect attack worth troubling about. It is notable also that in the case of string beans grown in this garden, the Mexican bean beetles, originally a pest with us, have entirely disappeared except for a return during a wet season when they attacked plants in spots where the water stood for long periods on the surface. In this isolated case, of course, the reasons were obvious—the saturated soil inhibited the entrance of all oxygen or nitrogen as well as the processes by which the plant normally fed itself from the available elements. The minerals were there in the soil but became temporarily almost wholly unavailable to the plants owing to poor drainage. The plants became sickly and the bean beetle returned to attack them. It was remarkable that the beetles did not attack the healthier bean plants on high well-drained ground in the same garden. It is also true that, under the poorly drained condition of the soils, the anaerobic bacteria which flourish in wet land gained the upper hand over the normal benevolent bacteria operating in good well-drained soil.

Of course, any gardener, from the backyard suburban gardener to the truck garden specialist, knows the preference of the insect for a sickly plant—that they will attack the sickliest plant first, then the next sickliest and so on all along a row. At Malabar we have been attempting to discover the relationship of given soil balances suited to given plants in their resistance to both disease and insects, and in some specific instances the results have been remarkable. The case of the bean beetle has already been cited. The case of celery and its resistance to mosaic blight, a universal pest of celery growers, is even more striking.

About three years ago we achieved in the "leafy" garden apparently a balance perfectly fitted to the health and resistance of celery, with the result that since then we have grown celery wholly free from blight without dusting or spraying *in soil into which we had deliberately turned back infected stems and the spores of mosaic blight for a period of six years.* Some twenty trace elements had been incorporated in the soil plus the establishment of fairly normal balances among the major elements—nitrogen, calcium, phosphorus and potash. As these balances became established, we grew celery which showed each year an increasing resistance to blight. Nitrogen turned out to be the key to total resistance. Despite the fact that the nitrogen content of the "leafy" garden was high, too high for any root crop and even for tomatoes, it was apparently still not high enough for the demands of perfectly healthy and vigorous celery, which is even more greedy of nitrogen than corn. When we added nitrate of soda or sulphate of ammonia, (both high nitrogen fertilizers) the celery became completely free of all mosaic blight without any dusting or spraying whatever.

At Malabar we do not pretend even to know a fraction of the reasons for the results obtained. Of one thing, however, we are persuaded—that when the proper balances for a given crop are established, that plant becomes almost totally resistant to disease and to attack from most insects. The one exception perhaps lies in the field of what might be described as "pest" insects like the locust or grasshopper, which appear in vast numbers migrating from some distant breeding ground with an appetite not for any given plants but for anything up to fence posts.

It is not impossible that there are other factors involved beside the balance factor. A clue in that direction in relation to insects has long been known to hothouse growers of carnations and chrysanthemums—that if minute quantities of selenium oxide in solution were applied to soils, the plants set up an immunity to attack from the commoner pests associated with greenhouse horticulture. Selenium is known, of course, to be a toxic element with the power, when used in sufficient quantities in soils, not only to poison insects but even animals and people feeding from vegetation produced on such high selenium-content soils. In other words, instead of spraying plants with poisons, it is necessary in the case of some plants merely to incorporate the poisons in the earth in which they grow and the plant will do the poisoning itself or at least render that plant highly or wholly unpalatable to the attacking insect. While the selenium process is effective in the case of ornamental flowering plants, and would doubtless be effective in the case of vegetables and forage crops as well, it is unsafe and impractical to use on food crops, either for animals or people, because of its toxic qualities.[5]

Much excellent and pioneer work on nutrition in relation to plants, animals and people has been done at Missouri University, and Leonard Haseman, Chairman of the Department of Entomology at that university, reports the results of research done at the University Experiment Station as follows:

With some of our most troublesome crop pests, there is a direct relation between the insect numbers and the soil fertility—*the less fertility, the more insects.* Our experience and studies at the Missouri Agricultural Station over the last several years have proved this.

In other words, the indications are that as we mine our soils by over-cropping, single cropping and by encouraging soil erosion, we are growing crops more and more deficient in some important nutrients. But these crops are increasingly attractive to insects. And the insects find them very good food for building still larger insect pest populations.[6]

One factor of especial interest arising from the research is that an abundance of nitrogen actually proved in many cases to be an insect

[5] The Battelle Metallurgical Institute of Columbus, Ohio, has done much notable work on trace elements in relation to the health and vigor of plants, animals and people. One of its staff, Frank A. Gilbert, is the author of the best and simplest book I know on the relation of minerals and balances. It is called *Mineral Nutrition of Plants and Animals* and is published by the University of Oklahoma Press, Norman, Oklahoma. Like *Hunger Signs in Crops* published by the National Fertilizer Association of Washington, D. C., this book should be among the reference books of any man who considers himself a farmer.

The trace elements used in the truck gardens at Malabar came from several sources: the Tennessee Corporation of Atlanta, Georgia, which markets a compound trace-element fertilizer under the trade name of Es-min-el; the Dow Chemical Company which puts out a product known as Sea Soil, which is the residue left from processes in which magnesium and chlorine and sodium are extracted for commercial purposes; and a product called Soilex, temporarily out of production. Many of the standard garden fertilizers such as Vigoro contain trace elements in considerable quantities, and more and more commercial fertilizer manufacturers are displaying a willingness to mix them into standard fertilizer combinations. Some are even putting on the market fertilizer containing the known and acceptably beneficial trace elements.

[6] From *Making a Better Living from Your Soil*, published by Successful Farming, Des Moines, Iowa.

[27]

repellent, a discovery which paralleled closely our own experience in freeing celery completely from mosaic blight by an increase in nitrogen fertilizer, even upon soils already heavy in nitrogen content. This information would follow closely the rise in insect pests over the country as a whole in ratio with the general depletion of organic material and, to a large extent, of available nitrogen as soils "wore out," and through lack of moisture even nitrogen commercial fertilizer became unavailable during the hot dry months.[7]

While the results above have been obtained and observed in an intensive fashion in the truck gardens, remarkable evidences of the same reactions have been obtained and observed in the fields themselves, especially in relation to alfalfa, one of the best of plants as soil testers because it shows deficiencies of all sorts as clearly as the spots on a child reveal measles or chicken pox.

During the first year of an alfalfa seeding and even well into the second year, our alfalfa plants have always shown signs of many deficiencies and have been subject to violent attack both from spittle bugs and leafhoppers. By the third year, however, the signs of deficiencies (save under occasional severe drought when through lack of moisture minerals may become temporarily unavailable) virtually disappear and with them disappear the spittle bugs and leafhoppers as well.

Frequently we have two strips of alfalfa side by side with not so much as a fence between them, the one strip showing deficiency signs and covered with spittle bugs and later in the season with leafhoppers. An inch away on the second strip there will be virtually no evidence of deficiencies and no spittle bugs or leafhoppers. The surface soil is the same. The seed is the same and the fertilizer and soil-fitting methods are the same on both strips. The difference, we find, is brought about entirely by the depth of root growth.

During the first year or two, the alfalfa, with a root depth of up to a maximum of eighteen inches, feeds on the top layer of soil which

[7] The author's own interest in the whole field of mineral balances and organic material in relation to insect control came about originally from the contrast between the vast difference in degree of insect attack in France and this country. He farmed and gardened in northern France for fifteen years without ever once having to use an insecticide, while during the early years at Malabar the gardens and fields were overrun by insect pests of all kinds—this despite the fact that the French soils had been cultivated for centuries longer than the soil of Ohio. The agriculture of France had during that long period been on the whole on a much higher level than that generally practiced in the United States. French agriculture has been persistently an agriculture of construction rather than destruction, in which all waste is returned to the earth from which it came.

had been farmed for more than a century to a depth of less than ten inches. Not only was this top layer depleted by a poor agriculture during most of that period, but the minerals had leached badly under constant row-crop cultivation to a lower level. On most of the land, insufficient fertilizer or none at all had been used over a long period, and the only fertilizer used, save for a small amount of barnyard manure, had been the standard chemical fertilizers containing only phosphorus, nitrogen and potash. The fields had never been limed, and even of the major elements, insufficient amounts had been replaced to replenish the drain from a constant row- and cash-crop agriculture. The result was that the top layer became deficient in almost every element and very nearly totally deficient in some of the trace elements since no effort whatever had ever been made to replace them. Even our own applications of fertilizer over considerable periods had not succeeded in raising the level of mineral fertility and balance to that of the deep rich gravel loam subsoil or of the original virgin topsoil. With these factors in mind it is easy to see why the alfalfa behaved as it did in relation to deficiency signs, disease and insect attack.

I repeat that the story was told entirely in terms of root depth. The first- and second-year alfalfa, growing and feeding in the top, shallow, depleted layer of soil, displayed all sorts of deficiencies including those of manganese and boron. It was sickly and therefore non-resistant to disease and to insects. Once its roots penetrated into the deeper level of our virgin glacial fertility, the signs of deficiencies disappeared altogether and with them disappeared all signs of disease or attack by insects. Thus the two strips side by side, and indeed with the leaves of one strip touching the leaves of the plants on the other, showed a remarkable difference. On the younger alfalfa, deficiency signs were everywhere in evidence, and there were literally millions of spittle bugs and leafhoppers. On the adjoining three-year-old strip in which the roots had attained a depth of fifteen feet or more into the virgin well-balanced glacial drift and abundant moisture, there were no signs of deficiencies whatever and it was difficult to find a single spittle bug or leafhopper.

These very strips have been seen by thousands of farmer visitors as well as by entomologists, agronomists, ecologists and medical research authorities.

On these fields we have never used any applications of trace elements with the exception of boron and of cross strips here and there where experimental applications of various trace-elements combinations were laid down in strips for observation. Within a couple of years, these experimental strips showed an improvement in the quality and resistance of the alfalfa plants, but after the second year, once the alfalfa roots had penetrated into the rich subsoils, the plants were vigorous and resistant on their own, even more so than the plants on the cross strips where the applications had been made.

Not all farms of course have recourse to the deep, well-balanced glacial subsoils that we have in a large part at Malabar. Many farmers on a different type of soil where the deep minerals were not so highly available have corrected the trace-element deficiencies in relation to alfalfa by direct application on the surface of the missing or depleted elements and have achieved the same results on the score of resistance to disease and insect attack. One notable example is Mr. Forrest Borders who, farming on the red clay soil near Bowling Green, Kentucky, has increased the yields and quality of the alfalfa on his excellent dairy farm by as much as 100 per cent. He reports the same diminishing attack from spittle bug and leafhopper to the point virtually of immunity.

The process of drawing on our rich glacial subsoils for the restoration of proper mineral balances in the worn-up topsoils is discussed in much greater detail in the chapter "Farming from Three to Twenty Feet Down."

One element in these balances which has been omitted thus far in the discussion is the element of moisture, which plays a large role since it determines to such a great degree the availability of all fertilizer, whether in the form of natural fertility or of bought chemical fertilizer. This is true whether there is too much or too little moisture. It has been omitted up to this point because, essentially, it is so closely related and involved with the factor of organic material.

Soils without organic material or low in organic content turn available water from the surface off the land or refuse to drain it off properly so that the water remains on the top level of the soil. Either condition upsets the whole of any carefully arranged mineral soil balance (as in the case of the string beans during a wet year). In the one case, the lack causes an *artificial* drought since the water runs off without ever penetrating very deeply and there is no sponge-like humus to retain it by preventing evaporation. On the other hand, the poor drainage in soils low in organic material creates a kind of *artificial* morass in which again, because of too much water, a whole carefully planned scheme of balances is nullified because the saturated condition of the soil inhibits all the natural processes by which the plant feeds itself. Thus abundant organic material works beneficially in two ways, to prevent both an artificial drought and an artificial swamp in which only very specialized plants can remain healthy or even survive. Cactus represents the extreme tolerance of land plant life toward shortage of moisture and bulrushes the extreme tolerance of land plants toward excess moisture, but I doubt that either cactus or bulrushes will ever prove very profitable crops for any farmer. Valuable plants, of course, vary greatly in their tolerances of moisture as they do toward mineral balances. The succulent and nutritious grasses and legumes of the Great Plains exhibit a great tolerance for seasonal droughts, and alfalfa will not tolerate poor drainage for even a few weeks.

Again none of this is simple, but neither is a good agriculture. But the farmer who makes some attempt at understanding and practicing these things will show not only gains in income but in the health and quality of his production and of his animals and family, producing at the same time great benefits for any who are fortunate enough to consume the products he raises, all the way from green peas to hamburger.

CHAPTER III

The New World in Agriculture (cont.)

Though our definition of the terms of academic freedom has been vague and some individuals and groups in our society have been suspicious of the principle, we have recognized generally that democracy rests upon a balance between faith and a search for truth, between scepticism and conviction, which can be attained only in the atmosphere of freedom and the courage which freedom alone can give.

— SARAH BLANDING, *President of Vassar College, in her Inaugural Address*

III. *The New World in Agriculture (cont.)*

THE relationship of soils and minerals to the health of animals and people becomes even more complicated than that of the relation of these things to the health and resistance of plants, but medical research, the work of specialized foundations, the observations of good farmers and the work of some state agricultural colleges is rapidly fixing a pattern which bears a close resemblance to that bearing upon plants. Indeed, the discoveries are coming so rapidly that they cannot quite become digested intellectually and brought into effective correlation in the whole of the picture. The question of the relation of these things to the health of animals and people is immensely complicated by other factors which, although slowly emerging, are not yet clearly understood either *per se* or in their relationship to one another. They are complicated by such elements as the functioning of the whole endocrine system and the specific purposes of the individual glands, by factors of metabolism, by the anti-biotics such as penicillin and the anti-bodies (serums, vaccines, etc.) as well as the amino acids, the vitamins, the enzymes and hormones, fungi and moulds and the animal secretions contained and operating in well-managed barnyard manures. The final and complete answers to many of the existing mysteries are fundamental and, scientifically speaking, exact, and no amount of quackery or short cuts or improvising or evasion will solve them.

In so far as the discussion goes in this chapter, it will be confined merely to a record of observation and the results of certain practices. We do not pretend to provide the whole of the explanation as to why, on a properly managed farm operating under the principles of a really modern agriculture, the rate of disease and infection is infinitely lowered in comparison to the average. We, as farmers, know *what works* by the results, and we do have theories and intimations of *why* some things work, but the full answer in terms of reasons is certainly unknown and is probably still fairly remote. In the chapter devoted to "The Chicken-Litter Story" something of the whole pattern begins to emerge with some semblance of clarity and coordination. At Malabar we only know that we have a remarkably low rate of sickness and infections among plants, animals and people.

Operating a herd of some fifty cows constantly on the milking string with annual turnover about three times as great, we have virtually a zero record on mastitis. Acetonemia and milk fever are unknown and brucellosis or Bang's disease seems to take care of itself even among those imported heifers which, when tested a few weeks after arriving on the farm, show positive or suspicious tests either from vaccination or from natural causes.

Brucellosis abortus or Bang's disease is a common plague of dairy and beef cattle operators and a serious one because of the great losses it incurs either in the abortion of potentially valuable calves or because in some states there is a compulsory slaughter law for any cow showing a positive test. The basis of this law is not very well founded since, as one eminent researcher has put it, "What we know about Bang's disease we could put on a calling card. What we do not know would fill the *Encyclopedia Britannica*." Thus far the treatment of Bang's has been entirely along the lines of vaccines or quackery, both based upon a colossal and almost universal lack of knowledge regarding the disease, and with little or no attention given until very recently to the deficiency aspects.

At Malabar we do not pretend to know any more about the Bang's mystery than anyone else, but to date we are inclined to believe that our experience, our observation and perhaps our knowledge is as good as those of the next fellow's. In that assertion I think we could get the backing of such top medical research authorities as the Rockefeller Foundation and Johns Hopkins University.

Speaking for myself, I am inclined to believe that Bang's provides a notable case in which our research has gone, like so much research, wholly down one alley, perhaps blind, in which most medical and veterinarian research has gone to the virtual exclusion of all other factors since the time of Pasteur. I mean by that simply an absorbing and overall concentration on: (1) curing illness after it is established instead of preventing it in the first place; (2) seeking a solution *only* in vaccines, inoculations, serums and disinfectants again to the exclusion of all else, and in particular to the total exclusion of the nutritional, metabolic, endocrine and deficiency factors. Much of the research has been along the lines of putting spectacles on cannibalistic hens.

Again speaking for myself—and my theory is possibly as good as the next man's—I am inclined to believe that probably most or perhaps all cows carry the germs or virus of Bang's disease just as every person carries within himself the germs of tuberculosis, and that bad conditions of housing or certain deficiencies of nutrition or dislocation of glandular reactions can bring the disease violently into the open and produce abortion and positive Bang's tests.

There are many factors which seem to indicate this. Almost any cow will show a suspicious and sometimes a positive reaction to tests immediately before or after calving. Outbreaks occur on given farms in given herds without spreading to adjoining farms although the cattle on the next farm may have been rubbing noses with the "infected" herd across the fence. This is exactly the contrary of hoof and mouth disease which is well established as a highly contagious disease. At Malabar we import heifers at breeding age from Wisconsin and Minnesota, and although they show clean tests before they are brought on the farm, about 15 per cent become reactors, either positive or suspicious, within the period of five weeks before we test them again. In some cases this has been caused by vaccination, in others it is a natural development. However within another four to five months, it is virtually impossible to obtain a positive test from any heifer. We no longer segregate the suspected animals and we use the same bulls on all of them indiscriminately. Since this particular dairy program has been in practice for four years, we have had no abortions save in the case of physical accidents which could be clearly traced. (1) The almost universal course in animals vaccinated for Bang's is an immediate reaction giving a positive or suspicious test. This could most reasonably be explained by the fact that the animal probably already had the disease and that the immediate battle between the anti-bodies in the serum and the elements of the disease already present in the animal produce an immediate violent reaction which results in a positive test. (2) It is notable that even the most earnest partisans of vaccination make no claim that vaccination produces an enduring immunity as do most effective vaccines and inoculations. Many advocates of vaccination do not even claim a solid immediate immunity to outbreaks of the disease. (3) This leaves the whole theory of contagion, infection and vaccination in an extremely hazy state, especially when coupled with the fact that even the coagulation tests for Bang's are extremely tricky and in the hands of an inexperienced or careless veterinarian can produce wildly inaccurate results. The only two animals which we sold from Malabar because they could not be cleared up were two heifers out of a group which we vaccinated as an experiment. This occurred several years ago before the nutritional pattern now in operation had been well established. Let me point out that at Malabar we cannot afford to take chances, for we sell whole Grade A milk, we deal in dairy cattle and could not sell one off the farm without a completely negative test, and about thirty-five people drink the whole raw milk from the dairy every day of the year. It might also be added that we test regularly for both Bang's and tuberculosis.

I do not know the whole answer but I have some ideas, and these have good backing from many sources. In the first place, the American Medical Association reports it as improbable that anyone gets undulant or Maltese fever (the equivalent in humans of *Brucellosis abortus*) from cow's milk or dairy products, and that it is probable that the infections cannot be picked up through cow's milk but only through contact or eating the meat of infected animals. This is extraordinary in view of the fact that most laws calling for the slaughter of Bang's positive cows place no prohibition on the sale of the meat from that cow, one more evidence of the extremely nebulous and tentative state of our knowledge concerning the disease. It also reports that very likely few people are subject at all to infection from the cattle variety of Brucellosis but become infected by the variety affecting hogs and goats. It is notable that very few researchers and no medical authorities ever take a very positive stand with regard to solving the mysteries of the disease.

The most reasonable approach, weighing all the elements, seems to indicate that Bang's disease in cattle is perhaps always present and that, while it is not in itself directly a deficiency disease, such for example as anemia or acetonemia, deficiencies of certain minerals and perhaps certain vitamins or the presence of anti-bodies in the vaccine cause a flare-up in the disease, usually throughout a whole herd because that particular herd is suffering from certain deficiencies of minerals which may not be lacking in the soils or in the feeding program of the adjoining farm where the herd remains immune even though rubbing noses across the fence with the "infected" herd.

Possibly the most reasonable cure or, what is more important, the most reasonable preventive course is to make certain that the animals have access to a wide range of minerals, including most of the trace elements, in chemical compounds which make them readily available to the physiological system of the animal. This has been a practice at Malabar Farm since the very beginning when, aware of the "worn-out" character of our soils and the deficiencies perfectly evident in the top shallow layer, we kept a box of minerals continually beside the salt box both in the barns and in the fields. During the first five to six years, until the program of deep-rooted grasses and legumes was well established, the cattle consumed as much of the mineral mixture as of salt, verifying our suspicion that the forage grown on the depleted top layer of our soil was deficient in the necessary minerals. From that period onward, as our grasses and legumes began to feed upon the deeper levels of the soils where the minerals existed in abundance, the consumption from the mineral boxes declined steadily. For the past two years the animals have consumed steadily less and less until at present many days go past when they never approach the source of mineral supplies at all, a fact which means that they are simply getting them, in the proper fashion and where they should, in organic form out of their forage itself rather than in inorganic form or in the form of the capsules and injections often used by veterinarians as a last-minute attempt at saving animals suffering from nutritional deficiencies or inability to breed.

Very often we find among the heifers imported from outside a craving for minerals which drives a considerable number of them to feed at the mineral boxes intensively for several days after their arrival at Malabar. Gradually, as they become accustomed to the new pasture and forage, their appetite for the minerals in inorganic form declines until they, too, scarcely ever visit the boxes. If one took the trouble to trace these animals back to the farms from which they came, it is almost certain we would discover that the land from which they were fed as calves would be badly depleted or deficient in minerals. There is a striking coincidence between those heifers showing positive or suspicious tests of Bang's disease and those which feed hungrily at the mineral boxes immediately after their arrival.

In support of the theory that Bang's disease is aggravated and made active by certain mineral deficiencies, much research of a serious nature has established the fact that the blood of positive test Bang's cows is invariably deficient in manganese and that a joint deficiency of manganese, cobalt and copper is probably one cause of outbreaks of the disease. It is significant that deficiencies of these three elements is well established as a principal cause of shy breeding in cattle, significant because these special deficiencies affect the same organs and glands which are affected by Bang's and its resulting abortions. Copper and cobalt are also established as the specifics in the cure of certain types of anemia in animals and people such as that occurring in the "droop neck" and "salt sickness" areas of Florida, South Georgia and Michigan, and the notable connection of cobalt with the newly discovered vitamin B–12, which is manufactured in the stomachs of the cow herself, should not be overlooked. Vitamin B–12 perhaps the most important of vitamins, cannot exist without minute traces of cobalt.[1]

The whole problem of *Brucellosis abortus* in cattle and of the corresponding undulant or Maltese fever in humans is immensely complicated and we by no means have the answers to its many complications and manifestations. The existing tests practiced upon cattle can be wildly inaccurate, and the diagnosis of undulant fever in people is extremely difficult and frequently misleading for the reason that undulant fever, supposedly at least, manifests itself in so many different fashions, disguising itself as a number of diseases apparently unrelated to it. There is no doubt that many illnesses are diagnosed as such when the illness is caused by some other "infection" or by nutritional and glandular disorders.

Quite recently the State College of the University of Missouri has undertaken research of a strictly scientific nature regarding certain trace elements in relation to the control and prevention of Bang's disease. The results, even in a short period, have been remarkable. In a report on *Brucella Infections and Their Possible Relation to Deficiency of Trace Elements in Soils, Plants and People*, by F. M. Pottenger, M.D., Ira Allison, M.D. and William A. Albrecht, Ph.D., the following quotation occurs:

In addition to the use of trace-element therapy for brucellosis in human beings, this same therapy now has been used for almost a year on a herd of dairy cattle with BRUCELLA infections, abortions, difficulties in breeding and other perplexing irregularities. During the year preceding the feeding of the trace elements to the cattle, there was a total calf crop of twenty viable calves from fifty-six cows. During the past year of trace-element therapy, each of fifty-two cows delivered a calf, including two abortions, and one injured fatally. These calves were larger at birth than those of previous years. Fewer irregularities have been experienced in getting the calves started. The number of services required for a single fecundation, either by artificial or natural insemination, has been significantly less during the past year, whereas previously it had mounted to as high as fourteen. The milk production per cow has increased to push up the average for the herd to a good record.

Soil treatments with the trace elements as fertilizers for the grazing

[1] For those interested in the subject, information can be obtained from J. F. Wischusen, Manganese Research Foundation, 1503 Lake Shore Drive, Cleveland, Ohio, as well as from Dr. Ira Allison, Columbia, Missouri, who has had notable success in alleviating and curing undulant fever in humans through a therapy based largely upon manganese, cobalt, copper and iodine.

crops also have been emphasized. As the herd started grazing on the treated soils, the previous therapy of trace elements was omitted and no irregularity was noted. However, the improved milk production was lowered when the trace elements were put into the feed again in addition to those coming with the forage. Here is a suggestion that it may be possible to overdose by adding trace-element salts to the feed if the soils already are providing these trace elements via the forages in the fields.

The blood picture of the herd also has been under observation. There was the suggestion in the second sampling (which was the first after the therapy was introduced) that the therapy had caused the blood picture to become worse. Later samplings, however, indicate its gradual improvement, with the positive reactions moving toward the lower dilutions used in the tests, with the number of negative animals increasing, and with the number of suspects decreasing.

The soil treatments using the trace elements are being applied in addition to a carefully planned program of soil-building of higher fertility by means of the major nutrient elements applied in accordance with suggestions coming from careful testing of the soils. While it has not yet been possible to demonstrate the efficiency of the treatments of the soil with trace elements as better nutrition for prevention of BRUCELLA infections of human beings, it is the hope that this treatment of the soil for better animal nutrition will prevent this ailment in the livestock. If it does, it will demonstrate the value of its coming by way of the soil as an initial step in the essential synthetic elaborations of these trace elements into organic combinations by microbes and plants in advance of those required in the intestinal tract, as is seemingly true for cobalt and copper. Such results with the livestock will be a step toward firmer belief and later proof that proper human nutrition also may mean the prevention of brucellosis and probably of other baffling ailments of mankind.[2]

This report follows very closely the experience at Malabar Farm over a ten-year period in relation to Bang's as well as to a whole range of other cattle diseases and to shy breeding on which the record today is very near to absolute minimum. As has been stated above, the good results have increased steadily and a noticeable improvement occurred when fertility and organic material was restored to the soils at Malabar and the rich variety of all elements and minerals lying deep in the good subsoils were made available to the animals by the deep-rooted legumes and grasses in the natural form of forage. This would follow the indications of the research at Missouri Uni-

[2] The animal-soil aspect of the experiment is under the direction of A. W. Klemme, Extension Professor of Soils, University of Missouri, and the blood testing under the direction of Andrew W. Uren, D.V.M., of the Veterinary Department, University of Missouri.

versity, and almost universally elsewhere—that both animals and people should get their minerals and vitamins in the form of good food rather than in the form of injections and capsules and pills. The rise in resistance to Bang's and mastitis in the herd at Malabar was followed in a parallel fashion by the complete disappearance of all foot infections and of lumpjaw, both of which had been notable in the herd during the early years before the land was restored and the circulation of minerals from sources deep in the soil was re-established. At Malabar we have probably had a longer record of observation regarding these discoveries than any farm or institution in the nation, and in a sense the laboratory has been very nearly a perfect one in which to observe the rise in organic and mineral fertility and its remarkable effect upon the health and resistance of cattle to disease in all forms. For in the beginning most of the land was as depleted in fertility as it is possible for land to become in this glaciated area and disease of all kinds tended to disappear in exact ratio with the restoration of organic material and mineral availability.

With regard to the vitamin B–12 content of milk, the experiments conducted at Missouri University have been to date nothing short of startling, the content rising on an average upon the feeding of trace elements from 0.006 content to a content of 0.043. This is interesting in view of the now firmly established fact that traces of cobalt are necessary to the creation of vitamin B–12. It also indicates a notable difference in the nutritional quality of milk from cows fed upon good forage from good soil as against those fed on poor forage from deficient soils and especially those fed from soils deficient in trace elements, notably cobalt, copper and manganese. As to the mineral side, cows will take calcium, at least, out of their very bones to put it into the milk for their calves, and thus it is the cow herself rather than the milk and its calcium content which suffers. As is well known, cows on calcium-deficient forage will take so much calcium out of their bones to supply the deficiency in their milk that their bones will break merely under the effort of getting up and down. This, of course, has nothing to do with the vitamin content of the milk which the cow cannot replenish out of her bones or physiological make-up. If the cobalt is absent, the cow cannot manufacture vitamin B–12 in her rumen, and not only will she fail to produce it in her milk but actually will herself become anemic and eventually cease eating and starve to death.[3]

[3] Much more detailed information can be obtained from Missouri University, Department of Agricultural Extension Work.

[42]

All of these discoveries and research have opened up whole new fields of speculation and have tended to put into a bad light many of the theories and practices existing in the past. They offer the strongest support to the new trend in medicine toward preventing animals and people from becoming ill rather than attempting to patch them up after the illness has occurred.

It is now well known that poliomyelitis, or infantile paralysis, was frequently diagnosed in the past as everything from a cold in the head to cholera morbis and that actually the great majority of adults now over thirty were once victims of the more harmless manifestations of polio wrongly diagnosed. This is the probable reason why polio, once commonly known as "infantile paralysis," attacks adults comparatively rarely although the records of such attacks have been on the increase for some time past. But again, polio is an extremely mysterious disease about which we still know almost nothing, and falls into the category of those mystery diseases which are attributed to viruses which we "cannot see under ordinary microscopic examination." This, of course, as has been suggested in a later chapter, provides an extremely convenient catch-all for what we do not as yet understand or have not solved. It is interesting to note that, among a small circle of reputable doctors familiar with polio, there has arisen a suspicion that polio may not be an infectious or contagious disease which is the result of germs or virus at all but is merely an alkaloid poison passed through the cow in the milk itself to the human, the theory being that, on poor pasture or during the hot dry season when pasture is short, the cow is forced to eat vegetation, often poisonous in character, which she would normally avoid and that she may herself remain more or less unaffected while passing the poison along through her milk.[4]

There are many facts which would tend to indicate the possibility of such a theory: (1) The most violent outbreaks of polio occur in the months of July, August and September when the weather is hot and dry and pasture is in short supply. (2) The most widespread outbreaks of polio occur in the Southern and Southeastern states where the pasture shortage is much more acute, during the hot dry months, than in other areas. (3) Most of these areas are notable for the presence in great quantities of the Jimson weed and cocklebur, both producers of alkaloid poisons and notably fatal to animals when eaten in sufficient quantities. (4) The symptoms of such alkaloid

[4] I believe that Dr. Marsh Pittsman of St. Louis, Mo., was the originator of this theory.

poisoning in animals, whether fatal or not, resemble very closely the symptoms of polio in humans, notably in staggering, paralysis and occasionally convulsions. (5) Alkaloid poisons attack with special virulence the nerve centers and notably the nerves affecting the lungs. (6) Polio is more prevalent in small towns and villages where the milk comes from a few sources in an area possibly affected by a drought which severely cuts down the growth of grasses and legumes and forces cattle to eat whatever they can find at hand. The milk going into the great cities comes from a vast number of dairy farms and is mixed together in a common pot so that any poison present is subject to a heavy dilution in which "infected" milk and clean milk is all mixed together. (7) There may be some connection between the prevalence of the disease among children and the fact that children and especially *American* children are fed great quantities of milk. It is notable that polio is infinitely rarer in Europe where the habit of milk drinking is comparatively limited and in Latin countries almost unknown among adults and rare with children. Northern Europe, where some milk is drunk, is rarely subject to the droughts or even the dry spells common during the summer months throughout the whole of the United States. Generally speaking, under the widespread well-balanced rainfall of North Europe, pasture is always green and abundant. Outbreaks of polio in Europe usually coincide with drought years, although they rarely attain anything like the virulence of the attacks which have become commonplace in the United States. (8) Polio shows a decided tendency for attacking the largest and healthiest children or, in other words, those accustomed to a steady and rich diet of milk, man's best-balanced food.

In all of this, as in the case of Bang's disease, the author makes no claim, certainly none of infallibility or even of any authority beyond that of observation and pragmatic results in the case of Bang's and mastitis, but the possibilities set forth above are of great interest principally in view of the whole chain of knowledge and information either released or co-ordinated in "The Chicken-Litter Story," and in view of the indications that research and theory have been so concentrated upon germs, viruses, vaccines and serums since the time of Pasteur that countless other elements in disease have been overlooked in some cases almost entirely. Each day brings some new discovery which completely pulls the carpet from under theories which in the past were accepted as facts. It is not impossible that many accepted theories of today may join the exploded superstitions of yesterday. The alkaloid poison polio theory may well be only one of those cases in which circumstance and an apparently impregnable chain of evidence merely throw the investigator off the real track.[5]

It might be noted in passing that "forage poisoning" among veterinarians is as convenient a catch-all for what they cannot diagnose or explain as is "virus" for specialists confronted by a baffling sickness. Today it is quite common to hear people who are virtually illiterate referring to their "virus." One of the commonest expressions among the uninformed today is "I had a virus," although they have no idea on earth what a "virus" is. Some, I am persuaded, regard it as some kind of animal, but it is a great help to puzzled or uninformed doctors.

Although good nutritionally balanced forage undoubtedly has much to do with the absence of mastitis in the dairy herds at Malabar, there are other factors involved as well. Under the dairy program at Malabar a considerable percentage of the cows in the milking parlor are heifers which are less susceptible to mastitis than older, heavier producing cows. There is also the factor of the loafing-shed-milking-parlor system of housing which keeps the cows off the cold concrete of stanchion barns where on a zero winter day the temperature can be as much as 40 degrees colder than on the loose straw above the accumulated manure of the loafing-shed. The pen-stabling system, which is approximately the same as the loafing-shed system, also prevents the joint injuries and the stepping on udders which lead to mastitis infections. In view of "The Chicken-Litter Story" told in Chapter V, there is a likelihood that fungi, moulds, benevolent bacteria and anti-biotics present in the accumulated straw and manure in the loafing-shed serve greatly to reduce or even eliminate the number of mastitis bacteria which can be very prevalent where manure is not allowed to accumulate and where disinfectants and daily fresh straw are in constant use as in a stanchion barn. It is also possible that, as the organic content of the soils at Malabar has been steadily increased, mastitis infections from the fields themselves have been greatly reduced by the presence of fungi, moulds and anti-biotics within the soil itself. There is also the fact that at Malabar there is a minimum of grain feeding or of any "forcing" of the cows to make production "records." Cows are treated as the ruminants they are, with stomachs designed to digest and utilize grasses and legumes and not heavy starchy grains. They are fed all the grass and legumes they will eat in the form of dry hay and silage.

[5] An excellent article on poisonous plants dangerous to cattle appeared in the Autumn, 1949, number of the *Farm Quarterly* (Cincinnati, Ohio). It is called "Poison in the Pasture," by Edward and Virginia Matson.

Both acetonemia and milk fever arise from deficiencies of calcium, sugar and possibly some trace elements. The fact that we no longer have any experience whatever with them tends to indicate that the herds at Malabar are obtaining the proper kind of nutritious forage and not suffering from deficiencies.

During the winter of 1949 an instance of the advantages of nutrition in humans turned up in an unexpected quarter without being sought. In the month of March one of the smaller boys on the farm had a cold and was kept home from school. When he protested, which seemed surprising, he said it was because up till then his attendance record since the opening of school had been perfect and he did not want to break it. I mentioned it to Al Bryner, the herdsman, and he said, "My three kids have had the same record." The remark led to a check among the thirteen small children and school-age teenagers on the farm to find that only two had missed a day at school in more than six months because of illness. Later in the season when an outbreak of measles and mumps went through the school and in some instances swept through whole families in the neighborhood, there were on the farm, among all the children, only two cases of measles and two of mumps which did not spread even within the families affected.

This record, obtained during a changeable and damp Ohio winter, is worth noting for whatever it is worth, but doubtless there is some connection between the record and the fact that all vegetables, fruits, meat, dairy and poultry products are raised upon well-managed and well-balanced soils, in which the good mineral balances of the deep rich subsoils have again been made available.

The great variations of the nutritional, mineral and vitamin values of vegetables and fruits according to the soils on which they are grown has long been established. In the case of lettuce, for example, the mineral and vitamin values can vary as much as 75 per cent. There is already a very large literature on this subject, and in the medical profession and in medical research there is an increasing preoccupation with the subject.[6]

[6] Among the many books besides Gilbert's *Mineral Nutrition of Plants and Animals*, I can recommend Sir Albert Howard's *The Soil and Health* (Devin-Adair), *Nutrition and Soil* by Dr. Lionel James Picton with an excellent introduction by Dr. Jonathan Forman (Devin-Adair), and *Soil, Food and Health*, published by The Friends of the Land, Columbus, Ohio. Also to be highly commended is an article by Firman E. Bear, Chairman of the Department of Agricultural Chemistry at the New Jersey State College of Agriculture. It is called "Regional Variations in the Mineral Composition of Vegetables" and appeared

At the very root of preventive rather than post-mortem or patent medicine lies the whole field of balanced and natural nutrition which in turn is based upon balanced, fertile and living soils. Certainly at Malabar the record of greater health and greater quality as well as greater quantity production and dollars and cents profits has followed almost exactly the line of increased organic material, better mineral balances and in general the establishment and maintenance of living, productive soils as against dead and inert ones.

To be sure, the whole of the New Agriculture, so far removed from the old conception that "anybody can farm," is not dependent alone upon the factors of soil fertility and the resulting good nutrition, although consideration and understanding of soil is its base. There are many other elements: (1) the new and more productive strains of crops developed by the plant breeder, although these are of no avail unless the quality and fertility of soils is the first consideration; (2) vast new improvements in machinery and mechanization, although these, too, are secondary to soils and may only serve to destroy the fertility of soils twice as fast as before if they are employed by the bad and ignorant farmer; (3) insecticides, dusts and chemical sprays which have done much to alleviate both disease and the attacks of insects, but on really good, living, productive soils the necessity for which is greatly reduced and in some cases obviated altogether; (4) developments both in irrigation and drainage, but again the need for both can be greatly reduced in most cases and sometimes be obviated by simple consideration for the element of organic material and its relation to the moisture factor, erosion and poor drainage.

One might list other factors which have contributed to the whole pattern of the New Agriculture, but essentially all of them are secondary to the establishment of good, living, productive and enduring soils which in themselves reduce the need for expensive and prolonged fitting operations, sprays and dusts, and even artificial measures related to both irrigation and drainage. Essentially the New Agriculture is concerned with soils and the related factors of good nutrition, health and optimum production both in quantity and

in the Autumn, 1948, number of The Land, a quarterly magazine published by The Friends of the Land, Columbus, Ohio.
   The subject has also been under constant study and report at the Nutrition Conference held annually in June at Ohio University, Athens, Ohio, under the auspices of The Friends of the Land and under the direction of Dr. Forman, President of the Society of Allergists and editor of the Ohio State Medical Journal. The proceedings can be obtained by writing to The Friends of the Land, Columbus, Ohio.

quality, and this is as it should be. In the past the soil was all too often overlooked or was made the victim of every sort of short cut and panacea to force artificial rather than natural fertility. Most of the time that path led to disaster in terms of erosion, poor drainage, lowered production in quantity and quality and seriously lowered nutritional value in foods both for beasts and humans. It might be said that in one sense medical research is following the same pattern as the New Agriculture—that is, turning to fundamentals through the emphasis upon preventive rather than curative or "patent" medicine, and in doing so is making remarkable discoveries similar and closely related to those made in the New Agriculture. Indeed, some of these discoveries are very nearly the same in both fields since they are so closely related and so interdependent.

But perhaps the most important element in the New Agriculture is the fashion in which it reaches out into almost every field of science and deals with the fundamentals of human existence whether it be war and peace, or economics, or health, or the problem of feeding a world which, at the old and the prevalent levels of agriculture, is overpopulated to the point of starvation in many areas. Men and nations, as the tragic happenings of the world reveal each day, are dependent primarily upon the soil for their very existence. In the era of the Industrial Revolution this truth came very near to being forgotten, but today both nations and people are being reminded of it day after day through bitter and tragic experience.

CHAPTER IV

The Golden Mine of Organics

Soil Fertility is the condition which results from the operation of Nature's round, from the orderly revolution of the wheel of life, from the adoption and the faithful execution of the first principle of agriculture—there must always be a perfect balance between the processes of growth and the processes of decay. The consequences of this condition are a living soil, abundant crops of good quality, and livestock which possess the bloom of health. The key to a fertile soil and a prosperous agriculture is humus.

—SIR ALBERT HOWARD,
An Agricultural Testament

Inorganic elements released in the soil by the decay of organic matter assembled there by Nature over long times are in seemingly better balance than we can prescribe them or supply them for food crop production. We are coming to appreciate organic matter in certain soil settings as Nature's gift of minerals broken down and their nutrients built into decomposable dynamic reserves.

—WILLIAM A. ALBRECHT,
Chairman of Soils Department,
Missouri University

IV. *The Golden Mine of Organics*

IN THE preceding chapters the value of organic material as a check upon both erosion and bad drainage has been developed at some length. This chapter is concerned principally with its value in increasing production per acre of any crop, in its power of making available to plants, animals and people the natural mineral fertility of the soil as well as the expensive commercial fertilizer purchased with good hard money by the farmer himself.

Any good farmer knows the capacity of organic material to retain moisture in the soil through a drought season. Any farmer knows that soil devoid or deficient in organic material drains badly or dries out rapidly under the action of the wind and sun and leaves a hard, caked surface. Any corn farmer knows that, no matter how much nitrogen fertilizer he applies to his crop of corn, it will "fire," display signs of nitrogen deficiencies and produce a miserable crop in hot dry weather if the soils to which he has applied the nitrogen are low in organic material.

Again and again in our Valley I have seen fields of corn with heavy nitrogen applications start off brilliantly during the June rains only to wither and shrivel under the hot sun and winds of July and August and yield a half or less of the prospect. The failure was less of rainfall than of organic material in such fields. The same failure cut the total corn production of the whole Middle Western corn belt many millions of bushels below the predicted yields of 1949. A long rainy season may produce good vigorous corn because all the fertilizer and especially the nitrogen is available to the crop, but it will produce late and soft corn if there is too much rain.

This element of too much rain rarely occurs in any of the areas within the Corn Belt. The problem is usually that of getting the corn through the dry hot weather of late July and August. If there is plenty of organic material in the soil, this problem rarely arises. The soil is open and porous and every drop of rain, even from a passing thundershower, enters the soil and stays there. Moreover, the very texture of soils high in organic material tends to preserve and hold the moisture and check evaporation. What holds true of corn holds true of every other crop. I have merely taken corn because its reaction to the presence or absence of moisture in the soil is perhaps more violent than that of any other crop and certainly more violent than the reaction of other grasses or of legumes.

Some eight years ago I made in Atlanta, Georgia, two statements: (1) that what General Sherman did to Georgia was insignificant in relation to what the Georgians had done to their own state by the constant unvaried growing of the cash crops, cotton and corn; (2) that the average Georgia farmer spending $100 on commercial fertilizer realized only about $10 or less of actual value. Because of the exceedingly low organic content of most Georgia soils after a couple of centuries of a single-crop, corn-and-cotton agriculture, the other $90 worth of commercial fertilizer very often went down the nearest stream along with the topsoil during the succeeding winter if the soil was left bare of cover crops.

The two statements made an uproar, the first among the old-time sentimental and shiftless Georgians or the ignorant small farmers and sharecroppers of whom there remain fewer and fewer each year. The second volume of uproar came from the National Fertilizer Association and finally developed into a controversy staged in the pages of their trade paper. Most fertilizer companies, even at so recent a period as the Thirties, still followed the "short-cut, cure-all" psychology and advocated practices which had wrecked millions of acres of our good American soils and ruined countless farmers in the past. It was based upon the premise that it was possible to farm with commercial fertilizer alone, and it ignored the whole relationship between the availability of fertilizers as well as natural soil fertility to organic material and moisture.

I was, falsely, accused of saying that commercial fertilizer had no value, which was and is certainly not true. I merely stated that its value was in almost direct ratio to the amount of organic material and consequently to the well-balanced moisture in any soil. During the controversy I pointed out that, in so far as the fertilizer business was concerned (as indeed in any business), a satisfied customer was a good and profitable customer, and that I had encountered many farmers who asserted that commercial fertilizer wasn't worth what they paid for it. Even the commercial fertilizer of today could certainly be improved, in most cases by a wider variety of analysis and by the addition in many cases of trace elements, but the statements of the farmers in question (invariably bad farmers who paid little heed or ignored the organic content of their soils) were certainly false. The fertilizer was certainly worth the money. The fault lay not with the fertilizer but with the farmer himself. Because of the poor natural conditions of his soils, he probably utilized at the most not more than 25 to 30 per cent of the value he had paid for. The residue, often enough, did not accumulate to be used by the crop the following year. Because his soil was like cement and his fields very often left bare during the winter, the residue of fertilizer at the end of the season merely washed away downstream into the sea. I also pointed out that some of the magazine advertising of the fertilizer companies could actually be the basis of a suit for using the mails to defraud since it read as if all one had to do in order to raise a bumper crop on an asphalt highway was to sprinkle a little commercial fertilizer on the surface and broadcast seed.

Whether or not the controversy had anything to do with a change in the subsequent policy of the fertilizer companies I do not know. I do know that in succeeding years the companies have published countless articles on the value of organic material and have hired top-quality agronomists with a profound belief in organic material as the basis of all sound agriculture to carry on a campaign for a better basic agriculture. At the national convention of the Association in 1949, a very considerable part of the program was given over to the virtues of organic material in relation to crop yields, farm prosperity and the consequent high availability and value of commercial fertilizer. The author, with whom the Association had carried on a lively feud only a few years earlier, was one of the principal speakers.

We have had much experience with the relation of organic material to drainage, erosion and the availability of natural fertility and commercial fertilizer at Malabar, perhaps as much as any farm in the nation. Not only did we take over both badly eroded and badly drained land but, with the exception of a relatively small acreage, we took over land almost totally devoid of organic material. On the high steep land the topsoil had all been eroded away down to the subsoil in which the organic content was an absolute zero. On the flat land where erosion did not take place to any great extent, the soil was so depleted through constant growing of cultivated cash-row crops, which burned out the organic material, that the organic content was virtually as low as that of the subsoils. The topsoils resembled cement with traces of acid rather than good productive living soil.

From the very beginning it was clear to anyone who knew anything about soils that no decent crops could be grown even with vast

[53]

expenditures on commercial fertilizer until the organic content of the soils had been vastly improved. It was toward this end that we directed our efforts and since then have increasingly developed them as the evidence told the story in terms of a total check on erosion and very nearly that on water loss and the return of good drainage in all our wet spots without the expense of putting a single tile into the ground.

The record of one field, described in *Malabar Farm*, was no less than startling. In the year we took over the Bailey place, one 10-acre field of wheat already planted on the farm was not worth harvesting and in the adjoining field there was not a cornshock as high as four feet. The wheat, which would have yielded less than 5 bushels to the acre, was mowed down and left in the field. Its potential organic content was worth more to us than the crop itself. Two years later on that same field we took 33 bushels of wheat to the acre and four years later we took 52 bushels.

We accomplished this gain, wisely, not by the use of vast quantities of commercial fertilizer (we used only 200 pounds to the acre of 3-12-12). We did it by the use of organic materials in as large quantities as possible, by lime, grasses and legumes and an enormous mine of old barnyard manure we found accumulated in the barns and barnyards of the farm. We even pulled down an old straw stack in the barnyard and allowed the cattle to run over it during the winter. By spring we had a heavy accumulation of well-rotted straw mixed with the droppings and the urine of the animals which went out upon the organically depleted fields.

All this organic material went on the field plus lime and commercial fertilizer. The lime was perhaps the key element, for it made it possible to raise that greatest friend of the farmer, the whole family of legumes. The lime not only sweetened the hard acid soils but permitted the legumes to grow and provide nitrogen and green manure, and the legumes in turn did many wonderful things below the surface in the soil which we could not see. The principal function of the commercial fertilizer was as a "starter" for the seedings of legumes and grasses. It provided the first step in the process of putting "dead" unproductive soil back into circulation.

In recent years, in the emphasis upon lime and ground limestone as the key to raising legumes and rich grasses, we have tended to overlook another great virtue of lime—its power as a catalyst to make other minerals and elements of fertility available to crops. In sour soils the minerals and elements go into chemical combination with aluminum and iron, the chief elements of most soils, in which form they become locked up and unavailable to crops. That is the reason why most sour land is regarded as poor land and many farms acquire the reputation of being "worn out" when they are not worn out at all. Their natural fertility is merely locked up in unavailable form. Nearly all elements and minerals have a much greater chemical affinity for calcium (lime) than they have for iron or aluminum, and once lime is added, they go into chemical compound with calcium (lime), in which form they are highly available to crops of almost every kind.

On the barren cement-like fields we took over at Malabar, there was beneath the worn-out, leached-out old topsoils and in the bare sub-soil below the top thin, leached-out layer, an almost inexhaustible supply of natural mineral fertility. The fields were not actually "worn out" at all. It was simply that the level of agriculture practiced by some of our predecessors was so low that the *native fertility* plus the fertility of the purchased commercial fertilizer was unavailable through the lack of lime and organic materials and moisture. More-over, with the possible exception of wheat and rye, no crop, even among the legumes, had been grown there in more than a century whose roots penetrated much deeper than eight or nine inches. That top layer, to a depth of nine to ten inches, was, it is true, sadly depleted and "worn out," but beneath that depth lay the rich primeval mineral fertility which had scarcely been touched.

The application of lime sweetened that top layer. The applications of fertilizer enriched it and the heavy application of barnyard manure not only enriched it chemically but added considerable quantities of organic material in addition to inoculating the whole of the field with fungi, moulds and bacteria which had been almost totally absent from the dry, caked soil for countless years.

Following the application of lime, barnyard manure and com-mercial fertilizers plus the mowed-down 5-bushel-an-acre wheat, the field was plowed very roughly and the surface disked until the grain drills could cover the seed of the new wheat seeding. The field went through the first winter very well, probably because it was rough plowed and loosely fitted. Under such a method, virtually no water was lost because the soil was left loose and open, with the old straw, the barnyard manure and the roots of wheat and weeds mixed *into* the soil to a depth of nine inches instead of being turned under, buried and then packed into a narrow layer by the subsequent fitting of the field. We had in effect made a kind of "sheet"-compost pile over the whole of the field in which, for the first time in many years, the elements of fertility, both natural and purchased, had a chance of actually reaching the crop.

In the spring we seeded 5 pounds of Hubam or annual sweet clover to the acre at the cheap price of 25 cents per pound. The seeding, owing to the lime and the first moisture the field had retained in years, came on well and by wheat harvest time was nine to ten inches high. The wheat was worth harvesting and actually produced in that first year about 18 bushels to the acre, which is the very low average production per acre of wheat over the whole of the United States. In harvesting it, we merely clipped the heads and allowed the straw to remain standing. From then until we began plowing again for wheat about the middle of September, the Hubam continued to grow until, at plowing time, the whole field was covered with a heavy mat of annual sweet clover growing up among the decaying straw. This provided many tons per acre of heavy organic material plus much nitrogen. The field was then given another dressing of the old barnyard manure and was roughly plowed, and again fitted only until the drills would cover for planting. It was a tough job but the results were worth it in the gains in yield accomplished.[1]

Undoubtedly the rough fitting of the fields in question, which we could not have avoided even if we had desired to, contributed much to the transformation of the rough straw and clover into good fertility-producing, moisture-holding humus, for the rough fitting left the soils open and aerated, permitting the oxygen and nitrogen, both from the air and the rain, to play their natural part in converting rough straw and stems into humus and soil. This could not happen on the old-fashioned, "clean" fitted field where the straw and green

[1] It should be remarked that plowing under the wheat straw *alone* would have done the soil very little good except in so far as its bulk, largely carbon, would have loosened up the soil and improved both drainage and water-holding capacity. Moreover, as any good farmer knows, the straw alone would have reduced the fertility for the crop which followed by actually robbing the soil of most of its nitrogen to feed the bacteria which in turn convert the coarse straw into soil and humus. Plowed in with a heavy crop of Hubam clover, however, the clover provided so much nitrogen in root, stem and in the nodules on the roots that the nitrogen provided an abundant diet for the bacteria and still left a plus rather than a minus of nitrogen for the crop of the succeeding year. In such fields it was possible to dig into the earth only three or four months after fitting and find virtually no trace whatever of the straw which had already turned into humus. On the other hand, in fields where straw or corn stover were turned over and buried by the old-fashioned clean plowing system, I have uncovered, while plowing, *whole* straw and cornstalks which had remained *mummified* in an underground, impacted layer and remained intact for as long as three to four years.

manure were first buried and then impacted in a thin layer by a fitting process which packed the clean soil above it into an airtight, almost cement-like layer, shutting out all nitrogen and oxygen from the air and often even from the rainfall. It is appalling how little we know as yet of the processes taking place *within* the soil.

The field was again seeded to wheat, and in the spring we made a second seeding of biennial sweet clover, red clover and alsike and timothy and got a good stand. This time we harvested both wheat and straw for it was clear from the beginning that we should have a good yield of wheat and that the straw that year would be heavy enough to damage the seeding. It went to the barns to be converted into manure. The wheat in this year, two years after we took over the field, yielded nearly 33 bushels to the acre over the crop we had inherited—a gain of 600 per cent after two years' treatment.

The following year the field went unplowed and we took a good crop of mixed timothy, alsike, sweet and red clover hay. In the fourth year the field was ripped up and fitted again, manured and fertilized with the usual 200 pounds of commercial fertilizer per acre and again seeded to wheat. This time into the wheat was sown in the following spring a permanent long-rotation seeding of alfalfa, brome grass and ladino clover, and at wheat harvest time we took off the field 53 bushels of wheat per acre.

This was not bad considering that the average yield per acre in the United States is 18 bushels and the Ohio average over a 25-year period is 22 bushels. And it was not bad coming off what five years earlier had been commonly considered "worn-out" land fit only for reforestation. It should be remarked that in the rehabilitation of that field, and indeed on all our land, we did not spend any more upon fertilizer than the average farmer raising wheat, and actually spent less than some who obtained lower yields. Primarily the job was done by lime, legumes and barnyard manure. The lime, without which no permanent prosperous agriculture may be maintained, was purchased. The commercial fertilizer was purchased. The seed for the legumes and grasses were purchased but actually the organic material which they produced was created without cost out of sunlight, air and water. The manure was a by-product, and an immensely valuable one, of livestock farming. Speaking, however, in terms of pure economics, the increased capital value of that land, based not upon inflationary values but upon the Federal Land Bank appraisal of the yield per acre, increased about three times, and during the process we had taken off the land two crops of wheat, one worth at that time $66 per acre and the other worth $104, in addition to the good hay that was removed. This struck us as pretty good business, and the process proved two important points: (1) that the soil was not "worn out" at all but had merely been so badly farmed that the fertility was unavailable; (2) that working *with* Nature, she will reward you.

We could not have achieved the same results with commercial fertilizer alone or anything approaching those results. At the end of that period the soil was nearer to truck garden soil in structure than to the cement we had taken over. The erosion and water-loss factor had decreased by approximately as much as 80 per cent, which was true also of the drainage factor in the occasional hollows and low spots of our rolling land.

The field remained in brome grass, alfalfa and ladino for four years, yielding us each year during that period a heavy cutting of grass silage, a heavy cutting of first-quality hay and forty days or more of the finest possible pasture. This represented to us a return of at least $150 to $175 per acre. During this period no commercial ferti-lizer was bought or applied but only two applications of barnyard manure. When plowed up and put to wheat at the end of this period we took off the field a little better than 60 bushels of wheat without any nitrogen fertilizer whatever. Indeed, we had too much nitrogen, produced by the legumes without cost, so that the wheat lodged badly in small areas. In the meanwhile we had brought up from our subsoils through the deep-rooted grasses and legumes many valuable minerals and elements which had been virtually depleted in the original top-soils that had been farmed for over a century only about eight to ten inches deep with a cash-crop agriculture in which during most of that period the organic factor had been ignored.

This particular field represented one of our best records, indeed, one so astonishing that we did our best to discover all the factors in-volved—a very difficult operation in the case of soil. However, one conviction came to us and to our agronomist friends—that we had not only utilized through organics and moisture the natural fertility of the soil but the fertilizer purchased by the farmer ahead of us which had never been available to his crops because of the lack of organic material, moisture and all the other elements including fungi, moulds, bacteria and earthworms which go with a truly living and productive soil in which the eternal cycle of birth, growth, death, decay and rebirth are present. We also had abundant evidence in comparable operations that we could not have achieved such a result if we had "clean" plowed and fitted the fields in the old-fashioned conventional and traditional manner. By rough plowing and disking in the sods and manure, straw and sweet clover we established a kind of "sheet"-composting process which gave us almost at once 100 per cent of the value of the organic material grown on the field or spread over it rather than about 30 per cent or even less which we would have achieved under the old "clean" overfitted methods of agriculture. Moreover, all of that material, even the high carbon content, had been quickly converted into humus and soil.

Not every type of soil would perhaps respond in so rapid and striking a fashion, yet thousands of acres put to grass and legumes in the Deep South under similar treatment have shown much the same response and the same rapid development as have indeed many thousands of acres in our own area in Ohio and in the older parts of the Middle West. Almost any soil that was once reasonably good and productive will show a similar response under the same pattern of treatment. Barring the factor of total erosion to bedrock, it is much more difficult to "wear out" soils than was once commonly supposed.

In the early years at Malabar when we were operating under a more or less conventional four- to five-year rotation and were forced to raise some silage corn, we achieved similar results even while growing corn three to four years in succession. The catch was, however, that every year after the corn was removed to fill the silo we grew heavy crops of green Balbo rye to be turned in the following year together with a heavy layer of barnyard manure. Here again, we did not depend on a year or two of grass sod to supply organic material but used both rye and manure every year and actually increased the organic content while growing a crop of corn taken off for silage every year.

The wheat and Hubam combination, leaving the straw with the sweet clover in it to be plowed in each year, we have used regularly for some fields, planting wheat and Hubam for four years in succession. The records of production on these fields ran on an average about 5 to 10 bushels the first year, around 20 the second, somewhere in the 30's during the third year and over 40 during the fourth. By that time the process could no longer be continued as the wheat straw became far too heavy to leave for plowing and the fields went into the regulation rotation of five years upward in brome grass, alfalfa and ladino clover and back into small grains for a year before being seeded back again to grasses and legumes.

On similar fields which grew silage corn during the first years we ran a rotation in successive years of corn and oats and devised a system

[59]

of getting two crops a year while still *gaining* great quantities of organic material. After the corn was removed for silage, the field was seeded to rye as a winter cover crop with a coating of manure provided during the winter. In the spring all of this was plowed in and the field seeded to oats. In the oats was made a seeding of rye grass and lespedeza, a valuable annual legume for forage and soil restoration. Although some lodging occurred in spots, we had good yields of oats and a crop of straw to be worked through the barns into manure. After the oats had been harvested, the rye grass and lespedeza came on and provided excellent and heavy late-summer and fall pasture with the animals spreading their own manure. The rye grass continued to grow and spread even during the winter months and by the following spring had created a heavy sod with the prodigious root system typical of rye grass to be plowed in. Thus in one year we incorporated into the old, eroded and depleted soil one heavy application of barnyard manure, with the root systems and green manure of the rye seeding, the root system of the oats and a heavy rye-grass-lespedeza sod.

Under the present program at Malabar, the use of rye grass, Hubam, Balbo rye and other crops employed purely for increasing organic material is no longer necessary. Under a program of heavy deep-rooted grasses and legumes, varied only with small grains, there is no loss of organic material but on the contrary a constant and steady gain each time a coating of barnyard manure or a heavy deep-rooted sod or both is turned into the soil. Since the fields are plowed and fitted only once every six or seven years and then only lightly, there is no loss of organic material through constant plowing and cultivation and only a heavy gain which has changed the texture and structure of much of the soil from something resembling cement to something closely resembling truck garden soil.

On the poorest, steepest and completely eroded hill land the trash mulch system of cultivation advocated by Faulkner in *Plowman's Folly* proved the solution for restoring productivity. The hills on the Bailey place provided the most striking example of the power of trash mulching in the prevention of erosion, the creation of organic material and topsoil, and the change, economically speaking, from zero production to high-quality grass and legume forage in terms of silage, hay and pasture.

These hills had been corned out, farmed out, pastured out, sheeped out and abandoned. On them there remained not a grain of topsoil, and the vegetation covering them consisted of broom sedge, poverty grass, wire grass, sorrel, golden rod, wild aster and here and there a sickly tuft of bluegrass. Even in the months of May and June the hills, seen across the Valley from the Big House, were brown. Roughly speaking the 100 acres covered by the hills would not feed five head of cattle adequately during the summer months alone. So terrible was the prospect that when I told Max Drake, first manager of the farm, that I had bought the Bailey place suddenly and unexpectedly, tears came into his eyes. The restoration of that land to even the lowest sort of production seemed hopeless. The hills were known widely as the poorest land in the township.

At that time there was little or no available information on how to bring back such land. The only answer we got was that it was fit only to be abandoned or at best to be reforested. We had to find the solution ourselves and find it the hard way. We made the first attempt at restoration on the lower part of the hills where the devastation was not quite so terrible, and followed the then accepted process of plowing, putting in a nurse crop of small grain and seeding into it grass and legumes in order to achieve some kind of a sod.

The results of this accepted process were a complete failure. The nurse crop (oats) did not give us back the seed. The plowed ground, even covered by a thin stand of oats, eroded badly and the seeding, washed about in the process of erosion, was not worth keeping. Indeed, it appeared that we had only made the hazard of erosion worse than it had been when the hills were covered with scraggly weeds. The 2 tons of limestone and the fertilizer we applied appeared to have been lost. At the end of the summer, merely to check further erosion during the winter, we disked the field and seeded it to rye. The catch was fair and the rye provided a good amount of binding material in the form of a fairly sturdy root system, but the crop was not worth harvesting and, instead of mowing it down, we decided to try the trash mulch system advocated in Faulkner's *Plowman's Folly*. By the time we got round to the task there was a fairly heavy crop of weed seedlings of every kind growing up in the rye.

We went into the field with Ferguson tillers and disks and literally chewed the rye stems and the weeds and the roots of both into the surface to a depth of about three or four inches, leaving the trash on top or at best worked in to a shallow depth. The operation required going over the field six or seven times to get the desired results. Into the resulting trash mulch we drilled a seeding of alfalfa, brome grass and ladino with from 300 to 400 pounds of 3-12-12 fertilizer. Then something almost miraculous happened. Virtually every seed germinated and within a few weeks the field was green with seedlings of alfalfa, brome grass and ladino. Even under heavy rains there was no erosion whatever, for the falling rain was absorbed by the trash mulch as a sponge might absorb it. Before long we discovered that the greenness came not only from the seedlings of the legumes and grasses we had planted but from a tremendous crop of the weeds, whose seeds, accumulating during the many years the fields had been abandoned, had suddenly germinated, probably through the agency of the unaccustomed moisture contained in the trash mulch. Stimulated by the lime and fertilizer, the weeds were rushing into a rank growth.

There seemed to be only one course to take and that was to mow the weeds before they matured and produced seed. This we did, twice during the summer, and that led to another interesting discovery—that the weeds, mowed down, were an actual benefit to the grass legume seeding because the additional mulch which they created served to keep the ground cool and moist during the hot dry months on those eroded hills exposed constantly to wind and sun. The following year the once weedy, barren hills were covered with a heavy sod, virtually weedless, of alfalfa, brome grass and ladino clover, giving us the equivalent of $90 to $100 an acre in high-quality forage in that first year. As the seeding thickened up during the succeeding years, the yields increased, and after two years the hills which would not feed five head of cattle adequately pastured seventy-five head upward on rotated pasture. Some of the fields also provided us with grass silage and hay as well.

The operation had, of course, been made in defiance of much good advice from interested people. We had been told that alfalfa thrived only on deep rich loam and that in order to get a stand it was necessary to lime the fields two or three years before attempting an alfalfa seeding. We had seen much volunteer alfalfa growing rankly along the roadside where the lime content was high and the soil merely clay and had observed the thick stands of sweet clover existing under the same conditions and so, taking this tip from Nature, we went ahead with our seeding. Once lime and a little fertilizer were provided, the alfalfa and the sweet clover grew rankly on those barren weed-covered hills devoid of even a grain of topsoil and over the same once barren hills we now have heavy seedings of valuable grass and legumes. The ground limestone was not applied two or three years earlier but only the autumn before seeding.

The legume grass sods not only provided us with a very valuable crop for four or five years but the once barren soil, which originally would not give us our seed back when planted to a nurse crop in the beginning, gave us yields of 35 bushels of wheat per acre and upward when they were plowed four or five years later. And of course in the meanwhile there was no erosion and no water loss. The plowed-in sods plus the droppings of the animals pastured on those fields and a dressing or two of barnyard manure over a four- to five-year period increased the organic content, which had been almost zero in the beginning, many thousands of times. And when the sods were plowed in, we did not, of course, clean plow and fit the fields but worked them up roughly with sod, manure, roots, etc., well but loosely worked into the soil itself. Under such conditions all rainfall was absorbed and no erosion occurred even before the small grains had established a good root system.

At the same time we were engaged in developing alfalfa, not as a pampered plant but as a pioneer on poor soils, Dr. H. L. Borst was at work on the same project at the Soil Conservation Service Station at Zanesville, Ohio, not many miles away. Over a period of seven years with two hundred seedings at the station and on eroded hills similar to our own on neighboring farms, he proved beyond any doubt the efficacy and value of alfalfa as a poor land crop not only to provide an immediate, good income but to build up the land to a point where it would raise good yields of other crops.

Probably no contribution to the farmers of Ohio and elsewhere within recent years has been greater than the widespread proof that alfalfa is not a pampered aristocrat but a real pioneer when trash mulch plantings are coupled with lime and a little fertilizer. Today many hundreds of thousands of acres of eroded land in rough hilly country in the Middle West and the South have been reclaimed and are being reclaimed by this system, but it took a long time to break the myths about alfalfa, the "green gold" of the legume crops. Much of what we had been taught before might have been described as a course in how not to grow alfalfa.

While we were trash mulching the barren Bailey hills, a local farm machinery dealer stopped by the roadside and asked what we were trying to do with those hills which had long been known as the poorest land in the township. When we told him that we were seeding alfalfa, he gave a loud laugh and said, "When you raise alfalfa on the Bailey hills I'll come down and pasture it off on my hands and knees." That was five years ago and for four of those years we have had excellent crops of alfalfa, sweet clover, ladino and brome grass covering the Bailey hills. We have sent him many invitations to come down and fulfill his promise but he has never come. Nor, might I add, have some of the experts who said it could never be done.

But the greatest satisfaction is being able to look across the Valley from the Big House and see hills which had been brown, weed-grown and abandoned for nearly a generation now a deep rich green at almost any time of the year with a heavy growth of grass and legumes coming up to the knees of the fat shining coated cattle who rarely tasted expensive grain.

During the past few years there has grown up, principally in England and in this country, an extreme school of thought which decries *all* chemical fertilizer as destructive both to soil and to health. It is as extreme as the old school, now utterly discredited, which taught that it was possible to farm and get good yields from the continuous use of chemical fertilizer alone. In our extensive experience at Malabar, we have been unable to go along with either school of thought the whole of the way for a number of reasons, some born of observation and some based merely upon available knowledge and common sense.

The truth, it seems to us at Malabar, lies somewhere in between the two schools. Certainly no abundant profitable agriculture can be maintained without organic material which is still the chief factor in the availability and value of both chemical fertilizer and the natural fertility of soils, yet it is not always possible to build up soils by the use alone of organic manures, green manures and composts made off those soils if they are, as in many cases, naturally deficient in certain elements.

Nature laid down her soils in a haphazard way and, except for two or three soil types, few of them contain perfect balance of minerals. Occasionally one encounters soils which are almost totally deficient in certain minerals or trace elements or in some rare cases these may exist in chemical combinations which make them unavailable to the plants growing on such soils. Some of the world's poorest soils are virgin soils because of their lacks and imbalances, and in such cases no amount of green manures or other natural fertilizers *grown on these soils and composted from them* will correct the deficiencies or imbalances. In some form or other the lacking minerals must be brought in from the outside, and the cheapest and quickest way is through the use of chemical fertilizers employed reasonably.

Even in soils high in organic materials, the use of chemical fertilizers can be abused and, as agronomists have discovered, the excessive use of chemical fertilizer even in one year's application can do

[64]

serious harm. The excessive use of commercial fertilizer over considerable periods can and inevitably will do great damage even in the
presence of abundant organic material. There is no doubt whatever
that excessive use of chemical fertilizers will destroy much of the
living factor in soils represented by earthworms, bacteria, fungi and
moulds and so in the end actually *convert* living soils into dead ones.
Moreover, it is possible to saturate soils with auxiliary and undesirable
elements such as sulphur to a point where the soils become not only
unbalanced botanically but actually poisonous to plant life and possibly even to animals and humans.

Many theories have been put forward and some research has been
done in the field of this soil "poisoning" through the use of chemical
fertilizer and its relation to various glandular and digestive derange-
ments and even to the mysteries of cancer, but it is difficult, I think,
to find anything very definite or wholly convincing one way or
another in this respect. It is hazardous however to take an uncompromising stand either for or against such theories because so little is
known, so little has been proven and the element of co-ordination
and cause and effect in such cases has been so little studied.

In our own experience at Malabar the reasonable use of commercial
fertilizers to the maximum of 400 pounds to the acre on the poorest
land have shown no visible bad effects either upon the earthworm
population or, in so far as it can be measured, upon the fungi, moulds
and bacteria which play so large a part in the rapid conversion of
crude organic material into humus and good friable soil. The population of earthworms has increased many thousands of times from the
evidence of the spade and the turned furrow as the organic content
of the soil has been built up and more and more barnyard and green
manure, upon which the earthworms feed, has been incorporated in
the soil or distributed over the surface.

At the same time that the purely organic school of agriculture has
been growing up, there have been made some fantastic claims by the
crank element regarding earthworms.

There can be no dispute whatever regarding the immense value
of the earthworm in converting thin dead soils or soils low in organic
material into living and productive soils, but since the earthworm,
like all living organisms, must eat to live and must also have a certain
level of moisture in order to survive, his existence and his propagation
becomes extremely unlikely in soils devoid of organic material, sparsely
covered by sickly vegetation and subject to drying and baking under
sun and wind. His survival is therefore closely related to green or

barnyard manures, mulches and decaying organic materials in general. The claims that a few earthworms introduced into a cement-like field covered by sparse vegetation will soon rehabilitate the entire field and create quantities of topsoil are as absurd as the old-fashioned chemical fertilizer advertisements which implied that a sprinkling of fertilizer on an asphalt pavement would produce bumper crops.

As a general rule, where there is abundant organic material, vegetation, green and animal manures, the earthworm will move in, establish himself and propagate at a tremendous rate. At Malabar we had a barren eroded clay bank originally uninhabited by a single earthworm which we mulched with chicken manure for four or five successive years to fortify and benefit the shrubs planted there. During that period the earthworm population increased from zero to hundreds of thousands of worms who moved in to feed upon the rich chicken manure and litter in which there was a considerable quantity of spilled grain. Within four to five years they had aerated the clay subsoil of the bank and helped to create rapidly a considerable coating of dark topsoil which deepens with each year. However, they were *attracted by the manure mulch and thrived upon it.* Had it not been there, together with the moisture it conserved as a mulch, any colony of worms we might have introduced into the area would simply have moved out or perished from lack of food and from impossible living conditions.

In general, where there is organic material there will be worms and where there is none there will be no worms. The damage done by chemical fertilizer to worms is in our own experience dependent upon its reasonable use in quantity and is in inverse ratio to the amount of humus and organic materials in the soil itself. In other words, where there is abundant organic material there will be abundant and conserved moisture which tends to nullify the "burning" effects of chemical fertilizer and its capacity to injure or kill the worms, the bacteria and the fungi and moulds.

Certainly, in our experience, commercial fertilizer is of great value in producing rapidly the abundant vegetation and the green manures and organic materials upon which the earthworm is dependent for its very existence. It is also extremely valuable as a "starter" in feeding and fortifying deep-rooted seedings made in depleted, abused and worn-out topsoils until their root systems become strong enough to penetrate to deeper levels where virgin fertility still exists. We should have been unable without commercial fertilizer to produce the heavy sods that now cover our once barren hills devoid of all topsoil. Once the seedlings were "started" with the aid of the fertilizer, the seedlings advanced in growth and strength as their roots penetrated more and more deeply into the lower levels of the earth. Without the fertilizer, and especially the nitrogen element even in the case of légumes, few of the seedlings would have survived at all under the adverse conditions existing in the eroded, bare and depleted soils of the hill land. Once the seedlings have reached the point of survival, it is conceivable that their continued health and production could be continued merely by the composting of manures and vegetation resulting from the forage grown, but it is no more than conceivable, because there is always a lag between the amount of minerals produced from the soil and the amount returned to it in the form of animal manure and urine, a lag which is represented by the amount of meat, bone, fat, milk, hair, wool, etc., carried off a given field even in which no cash crop is grown at all and the land is used merely as pasture for livestock.

Moreover, the only certain way in which to maintain constantly the fertility while removing either cash crops or crops of milk, meat, etc., would be dependent upon the question of whether the minerals brought up from deep underground reserves to the surface became available rapidly enough to satisfy the full needs of the plants growing upon the surface. At Malabar we have indications that this might be true in the case of certain very deep and abundantly rooted plants such as alfalfa and sweet clover, and in deep gravel loam subsoils, but, as is pointed out in the chapter, "Farming from Three to Twenty Feet Down," we are by no means certain that this is true and except in test plots we are taking no chances. We use barnyard manure abundantly and, when the supply runs out, supplement it with chemical fertilizer in reasonable quantities.

There are one or two other factors involved in the organic-chemical fertilizer dispute. The first at least is purely economic and related in particular to the case of the farmer working to restore to fertility badly abused farm land of high *potential* fertility. The average farmer cannot sit about waiting to build up his land to profitable yields by the slow process of composting alone. In most cases he must obtain as quickly as possible yields which will pay for his taxes and interest, purchase his seed, fertilizer and equipment and still show at least some small degree of profit. Here is where chemical fertilizer can hasten the process in two ways: (1) by quickly providing him with reasonable yields; (2) by enabling him to produce rapidly considerable quantities of green manures and even barnyard manures (by increasing the livestock carrying capacity of his land along with his income) which he can turn back into the depleted soils to raise their quality and production.

Not long ago I had a letter from a friend who belongs to the rigid, uncompromising "organic" school. He wrote that he was very proud of the record he had made in raising wheat production by organic methods alone. He boasted that he had raised his production from an extremely low level, after seven years of composting and the use of organic materials, to a production of 20 bushels to the acre. As a matter of simple economics the average farmer would be in the hands of the sheriff if he failed to do a great deal better than that in a long period of seven years.

Our own record, in the raising of wheat production from 5 to 33 to 52 bushels per acre in five years has been set forth elsewhere in this chapter. We could not have achieved it without the use of commercial fertilizer both because of its direct and immediate effect and even more importantly because it aided us in raising much greater quantities of green manure and heavy sods to be turned back into the soil to raise the level of organic material. This is, of course, simply another evidence of the efficacy of chemical fertilizer as a "starter" in the initial kick-off operation on poor land.

There also arises the question of labor and machinery expense in composting all materials on a good-sized commercial farm. While machinery improvements are rapidly being made which are reducing very greatly the costs of the whole business of composting, the task still remains an expensive operation and in some senses an "extra" one which must be counted in on the cost side of any farm economy.

At Malabar we have consistently practiced what Sir Albert Howard calls "sheet"-composting. All the manure is kept under cover in the feeding sheds, even in the loafing-shed of the dairy barns, for a period of two to three months. During this period no fertility is lost either by "burning" or by leaching away and in the meanwhile the benevolent bacteria count has increased millions of times, the fungi and moulds have flourished and the carbon content of the bedding has been broken down by the action of all of these elements plus the action of the glandular secretions and various acids, and the raw sawdust, straw or shavings have been converted into the finest kind of highly available fertilizer.

After two or three months, this rich, moist, trampled manure is either spread over the meadows and pastures or over other fields where it is worked into the soils by disk, tiller or Graham plow to a depth of eight to nine inches. As meadows and pastures are always the base of a new planting of small grains, the soil also contains a great amount of roots and other organic material. This, together with the already half-disintegrated manure, is mixed *into* the earth and actually makes a compost heap eight to nine inches deep covering the whole of the field. It is in this shallow compost heap that our crops other than grass and legumes are constantly grown with an immense increase of moisture and highly available fertility. In this simple fashion the composting is actually achieved without any particular elaboration of the natural processes and without any "extra" work. In more recent operations in which the Ferguson tiller or the Graham plow is used we have been able to increase the compost depth to as much as eighteen inches.

In soils prepared in this fashion the growth of benevolent bacteria, fungi and moulds is enormously encouraged and increased to the benefit of the health and vigor of all plants and to the detriment of disease germs that may be present in the soils themselves. Moreover, the fungi, moulds and bacteria promote the mycorhizal action, which is only an elaborate way of saying that the bacteria, moulds and fungi actually feed the plants by translating mineral fertility into a highly available form and passing it on to them.

There has been too little research on this phase of living soils but in *Soil and Health* Sir Albert Howard produces many proofs, and in our own mulched vegetable gardens we have found again and again beneath the moist decaying mulch, fungi whose long filaments were actually attached to the fine hair-like roots of vegetables, apparently engaged in the process of "feeding" them. Sir Albert Howard has produced photographs of the same phenomenon. It is well-known that certain species of plants have certain special moulds or fungi which share a special affinity with them and work in co-operation. One evidence of this, known to any American farmer in the Middle West and to farmers in France, is the fact that one species of the delicious morel fungus is found almost invariably only under apple and ash trees. We have observed that when we destroyed an old orchard where these fungi had existed in abundance, the fungi failed completely to reappear once the apple trees had been destroyed. The same kind of affinity appears to exist between truffles and beech and oak trees in Europe.

The immense values of organic material to productive soils is indisputable and it is safe to say that, outside truck gardens, the greatest and most serious deficiency in the soils of the United States today is organic material. It is also probably the greatest limiting factor in high continued production and in the prosperity of the ordinary farmer. But again, it is not the whole answer to balanced and living productive soils but only a part of the whole pattern. Among my truck-gardening and hothouse-growing friends there have been many instances where, in their efforts to maintain the high organic content so valuable to them, they have overdone the process and produced soils that contained too much organic material, turned sour and produced a sickening odor which could not possibly be given off by any healthy balanced soil. There can scarcely be enough propaganda made in this country in behalf of organic material. On the other hand, exaggerated claims made for organic material to the exclusion of all other factors may prove unreasonable, uneconomic and unsound and lead many a farmer into difficulties.

It has been said many times and truthfully that it may require in the neighborhood of ten thousand years to build an inch of topsoil. Few people however add that man can build topsoils in a tiny fraction of that time merely by using the methods of Nature herself but by speeding them up many thousands of times. Nature built topsoils by laying down on the surface of the earth a fragile leaf or a delicate blade of grass one at a time over immensely long periods. She could not apply concentrations of animal manures. She could not grow crops of heavy-rooted rye and rye grass or heavy sods of legumes and grasses to be turned into the soil every year or two. She did not have the machinery which made it possible to incorporate these great quantities of organic material *into* the top few inches of the subsoils through the process of "sheet"-composting. She had only the power, described in the chapter "Farming from Three to Twenty Feet Down," of bringing up the almost inexhaustible supply of deep layer minerals and transforming them into organic form of leaves, roots, twigs, stems and blades of grass to be deposited on the surface. We, too, have that power by the use of the deep-rooted grasses and legumes. The minerals from deep down in organic form can be turned into the soil along with the vast bulk of green manures which come out of sunlight, air and water. The mineral drain on the soil itself in such a process represents a loss of considerably less than 5 per cent of the mineral content of the deeper soils and none whatever when no crop is taken from a field and the *whole* is turned back into the soil. One has merely made a great gain not only by incorporating the bulk of the green manure in the soil but also by converting inorganic minerals into organic form in which they are highly available to any succeeding crop.

With modern machinery and by knowledge and experiment and

intensive action we have been able to create on our gravel loam sub-soils from seven to eight inches of truck garden topsoil within as many years and brought back into high production some hundreds of acres most of which had been abandoned as "worn-out" useless land. I am not at all certain that these "created" topsoils are not better than much of the muck land in use for truck gardening since, on the muck land which is largely deficient, a great variety of minerals must be incorporated before the muck soil produces either healthy or nutritious vegetables. At Malabar we secure these minerals in immense variety out of our own glacial gravel loam subsoils. The deep-rooted legumes and grasses do it for us, bringing them up from deep down at the same time that they are providing us with immense quantities of organic materials formed out of sunlight, air and water. And the soil formed becomes a living and productive soil because it is filled with the fungi, moulds, bacteria, worms and moisture which all go to provide healthy plants and animals and people.

As in the case of lime and the neglect in emphasizing its catalyzing properties, so we have often overlooked perhaps the greatest and most valuable quality and characteristic of nearly all legumes—that they are one of the few families of plants which can grow and will grow even lustily in soils totally devoid of organic material and at the same time provide a valuable crop and create organic material for us in great quantities. One cannot grow corn in soils low in organics or even wheat or oats or barley with profitable yields, but one can grow legumes abundantly provided the mineral balance exists or is established. And the legumes have as well the great and, to the farmer, precious power of producing great quantities of highly available nitrogen out of the air itself without any cost to the farmer. Moreover, there are other powers possessed by legumes and in particular by alfalfa and the sweet clovers which have not been sufficiently investigated. Among these there is, without much question, their power to break down gravel and coarse soil particles into consistencies and chemical combinations in which their mineral fertility becomes available not only to the legumes themselves but to other plants and notably the grasses which grow in affinity with them after a very short period of time.

In our own gravel pit, cut out of a field which grows bumper crops of alfalfa, brome grass, ladino and even bluegrass on almost pure gravel with little or no topsoil and very little clay or loam content, we are able to watch closely the under-surface behavior of both legumes and grasses for we are constantly taking out gravel and with each fresh cut opening up new underground laboratories for inspection. Alfalfa roots fifteen to sixteen feet deep are common, and in every case, as the gravel is cut away and the fresh alfalfa roots left bare, we find that the long coarse main root is invariably surrounded by a column of real topsoil from a tenth to a quarter of an inch in thickness which has obviously been created out of the rather coarse gravel by some chemical action coming from the plant itself. There is nothing very remarkable about this because to a great extent it is merely a part of the action by which all rocks are broken down, beginning with the lichen on a bare granite boulder. There is every reason to suppose that all vegetation is constantly at work creating topsoil not only by the deposit of leaves, stems and other vegetation on the surface but by the action of underground roots as well.

As I have said elsewhere, we are all likely to overlook what goes on underground simply because we cannot see it, and in the past most of the research has been confined to inorganic tests which merely tell the actual mineral or elemental content of soils and all too often fail to reveal even whether the mineral content is in available form. The whole elemental and cosmic world of the fungi, moulds, bacteria, worm and root actions in breaking down and making available to plants (and consequently to animals and people) the essential minerals and elements has been very largely neglected or wholly overlooked save by a few men like Sir Albert Howard. The relation of all these elements to fertility and health in plants and consequently in animals and people, and even to heavy production, is only partly understood. The results, however, are accepted by the good farmer who knows his soil and what makes it productive without knowing why certain beneficial reactions, by which he swears out of long experience, take place and produce a given set of beneficial results.

The whole field of the capacity of these elements, including the roots of legumes, to create abundant and healthy nutritious crops and soil, is still largely unexplored. The mere fact that they have not been examined is not a sufficient reason for asserting that this capacity does not exist or that the root systems of most plants but particularly of the legumes do not possess certain immensely valuable powers of disintegrating and making available minerals contained in rock and rock particles. Both farmers and agronomists are likely to make the mistake of believing that the bulk or indeed the whole of the organic material derived from cover-crops or green manures comes from the visible green growth appearing above the surface of the earth, simply because *that is all they see.* In the case of many plants, particularly wheat, rye and the rye grasses, an even greater bulk of organic material and a higher mineral content comes not from the stems and leaves but from the roots—a fact which can be easily proven merely by observation.

One of the cases cited frequently in refutation of the strictly organic, compost school of farming is the test plot at Roehampton in England, where good crops of wheat have been grown every year for three generations by the use of chemical fertilizer alone. Certainly, the case is faulty as proof of the powers of chemical fertilizers alone for the simple reason that wheat, like many grasses, produces great quantities of roots which replenish the soil organically. One wheat plant in the dry plains of Saskatchewan was found to have a huge hair-like root system which virtually filled a cubic foot of earth. This, of course, is a prodigious amount of organic material, much of it developed deep in the soil. It is undoubtedly this factor which permits the growing of wheat as a single crop year after year in the Great Plains area without serious organic loss, although even there yields in many areas show a steadily declining production.

The Roehampton test plots merely prove that wheat can be grown year after year without replenishing organic material and with the use of chemical fertilizer alone. The tests are of little value in relation to other crops such, for example, as corn or maize, also grasses, but with a coarse, woody, shallow root system. Had corn or sugar beets been used on the Roehampton plots, the story would have been altogether different, with steadily declining yields to the point of disaster. Moreover, no records have been kept concerning the nutritional value of the wheat grown on such plots in terms both of minerals and of proteins. In our own Corn Belt, the protein content of corn has been steadily declining. By some authorities this decline is attributed to the widespread use of hybrid corn but it is much more likely that it arises from the gradual mineral depletion of these soils, and particularly from the decrease in availability of minerals and fertilizer closely linked to the steady depletion of organic material.

Not the least valuable factor in the presence of abundant organic materials in soils is the aeration of soils which they make possible. The effects of nitrogen and oxygen deep in the soil have been very little studied and very little information is available, but any good farmer or vegetable grower knows the vastly greater root growth and the greatly increased growing vigor of plants in "loose" soils abundant in organic material where the air itself is able to penetrate and even under certain conditions to circulate through the soil itself. The hunger of roots for air has been proven many times by the intensive root growth occurring around the edges of a natural or artificially

[73]

constructed "hole" in the earth; and often enough, in exploring the soil in a field or a vegetable garden, I have found the same vigorous even exaggerated growth of fine roots surrounding a "pocket" of air fairly deep below the surface. It is perfectly evident, I think, that soils fitted over and over again and packed tight by the constant passage on the surface of tractors and machinery will become impervious not only to rainfall but to the air itself and consequently produce dwarfed root systems and badly decreased yields.

This overfitting of fields to make them look "pretty" has undoubtedly contributed much to the depletion of organic material and immensely to the factor of soil erosion. As a Texas friend observed recently, "Pretty farming in America has cost the nation hundreds of millions of dollars and ruined many a farmer." The change from "pretty" farming to scientifically intelligent "rough" farming, and its beneficial effect upon wind and water erosion in the Great Plains area, has been dealt with elsewhere in this book. Suffice it to say that our experience at Malabar has led us steadily away from the "pretty" school of tillage into the rough and natural school, until today we fit any field as little rather than as much as possible and keep off every field as much as possible with machinery. The gains have been startling not only in the immediately augmented yields but even more so in the long-range building of soil texture, organic materials and the general tilth of the soils themselves.

One factor little understood is the affinity of certain plants which, in growing together, augment the vigor and productive fertility of each other. In Nature there is no such thing as a single crop; indeed, Nature abhors the single crop and the single crop has been the most destructive single influence in American agriculture. The evils of single-cropping extend even into the area of legumes, which undoubtedly flourish better and provide more and better forage when they are grown with grasses. In many parts of the country, the persistent single-cropping even of alfalfa has undoubtedly led to a certain deterioration of the soil structure and even to a decline in the organic content of soils since the alfalfa root, while deep-growing and vigorous, is coarse and woody and lacks the immense organic contribution made by the hair-like roots of grasses, in particular the deep-rooted grasses.

In our own practices we have long since ceased to grow alfalfa as a single crop. It is always mixed with the short hair-rooted ladino and with a variety of such deep and abundantly rooted grasses as the fescue, orchard and brome grasses. Not only does the mixture give us much greater yields of forage but it constantly makes enormous contributions of organic material through the annual death of most of the hair-like roots of the grasses which each year set forth new growths of hair-like roots. And there is the ultimate test provided by the cattle themselves which show a much greater preference for mixtures of grasses and legumes than for the finest pure alfalfa. Many times I have observed the cattle turning away from racks filled with the finest quality of alfalfa hay to devour the oats straw used for their bedding. I do not believe that the element of "wanting a change" is the whole factor. The nitrogen from the alfalfa roots undoubtedly increases the protein content of the hay, but it is not wholly foolish to assume that the grasses and their roots make a contribution as well to the palatability of the alfalfa. The cattle do not turn away from the hay made of mixed grasses and legumes to devour the oats straw but merely ignore it. Of course, the growth and tenderness and palatability of the grasses growing in green condition in the fields is greatly increased by the abundant nitrogen supply coming from fields impregnated with nitrogen from the alfalfa either from past accumulations provided by plowing in the legumes or possibly even from the existing alfalfa growing alongside the grasses.

Certain affinities are notable in given crops, among them the striking affinity of vetch and rye which when sowed together not only produce an astonishingly vigorous growth but a high palatability and nutrition as feed. In our once rich Great Plains grazing areas the heavy turfs once contained a mixture of grasses and legumes, but under overgrazing and burning over, virtually all the legumes have since disappeared, and on ranges where once, according to the old-timers, the grasses were so high as to hide the cattle, there is today little but a stunted, scattered growth of grasses with a nutritional value well below that of the rank-growing grasses which once consorted with the now exterminated legumes.

We have still only scratched the surface of the knowledge concerning what makes good, living and productive soils capable of growing foods which are wholly nutritious in the vitamin and mineral content. We are, indeed, only beginning to discover the relationship between certain minerals and certain vitamins as in the case of the direct relationship of cobalt to the production of vitamin B–12, one of the most important of vitamins, and the consequent relationship to anemia and the shy breeding of animals. The man who says we have learned all there is to know is merely a fool, as is any man who thinks he knows *all* the answers about any one phase of the universe. It is just possible that we need a little more inspiration, a little more imagination, perhaps even a little more religious faith in the Darwin-Huxley manner, regarding soils and agriculture than we have had in the past. It is out of such elements that new understanding of an immensely complex universe finally arrives. It is probable that even a kind of mysticism is necessary, for pure and real science is no more than the rationalization and cold analysis of instinctive mysticism.

Very often the good farmer, who is a keen observer, comes far nearer to an answer purely through pragmatic experience than the man with the test tube. Peasants in Northern and Central Europe have known for many centuries the practice of a good agriculture. They knew by observation and by result many things. They knew that a poultice of fresh cow manure was good for an open cut and that a poultice of mouldy bread would sometimes cure a stubborn ulcer. Actually they were dealing with vitamin B–12 and penicillin centuries before either was discovered as such or before either was given a name. The sacredness of the cow in the Hindu religion very likely arose from the virtues of the animal itself, not only because she provided milk and butter which play a large part in Hindu nutrition, but because her dung and her urine both had curative and fertility properties of great importance.

In the part of France where I farmed and gardened during fifteen years, the land has been farmed for fourteen hundred years, most of that time in wheat, for it has long been the breadbasket of France and exported wheat to other countries, and the average yields of wheat in that area are today 60 bushels to an acre as against an American average of 16 bushels on soils many of them farmed less than a century. But the Frenchman long ago knew the miracles that could be worked by organic materials, and none are wasted—neither the straw, the twigs, the weeds, the contents of the privy or, in the case of those ancient and abundant wheat fields, even the garbage and old shirts and underwear cast away by the city of Paris. All is hauled out into the fields and plowed into the good earth from which it all came.

Not all the virtues of organic materials are connected with fertility alone. It is now well known that all the miraculous anti-biotics such as penicillin and streptomycin are the products originally of moulds or fungi, and it is well-known that all anti-biotics discovered up to date save penicillin come directly from the fungi and moulds of soils high in organic content and preferably soils fertilized heavily with barnyard manure. Such soils, it seems highly likely, would be much less likely than poor soils, low or deficient in organics, to carry the malignant germs of disease against which the anti-biotics and their parents, the fungi and moulds, are such bitter enemies and destroyers.

On our own farm the record of diseases ranging from foot rot to mastitis has declined steadily to a virtual zero as the organic content and consequently the fertility and the high quality of mineral nutrition has improved. It is likely that the increased quality of mineral nutrition not only has checked diseases arising from deficiencies but that the count of malignant bacteria within the soils has decreased enormously as the fungi and moulds and benevolent bacteria increased. In other words, the same process taking place in the old chicken litter (described in "The Chicken-Litter Story") takes place as well in the manure of the loafing-shed and in the soils themselves once the organic content is increased and the fungi and moulds and benevolent bacteria have a chance to multiply vastly and perform their natural functions both in making minerals available and in killing off the bacteria of disease.

During both world wars American soldiers going into Northern France were every one scrupulously vaccinated against tetanus, sometimes at the risk of serious discomfort and even illness, on the assumption that Northern France was an area reeking with tetanus germs. I have never been able to uncover the research, if any, upon which this assumption was founded, but I farmed and gardened in that exact area and lived in intimate contact with the farmers, market gardeners and field laborers of the region for fifteen years. In all that period I never heard of a case of tetanus. Certainly I experienced innumerable cuts and scratches while actually working in those soils and, not being one to fuss, never even applied so much as a dab of iodine. Certainly I did not acquire tetanus or any other infection of any kind. Under such circumstances the fact that the record of tetanus in both American armies was extremely low, and sometimes perhaps wrongly diagnosed, proved nothing but that, unless they were very different in some biological fashion from the average French farmer or myself, they would not have contracted tetanus even *without* any inoculation. Indeed, it begins to appear likely, in the light of recent revolutionary discoveries, that those well-tended French soils, high in organic and manure content, are much less likely to contain the germs of tetanus infections than the newer soils of our poorer agricultural areas in the United States. (It was part of the legend that the French soils contained tetanus infections *because* they were *old* soils, a theory wholly contradictory in the light of recent discoveries.) In an infection as violent and marked and rare and as often fatal as tetanus it is highly unlikely that the French farmer or peasant has set up any immunity either by himself or through inheritance.

Neither should it be overlooked that many diseases such as beriberi, pellagra and the "droop neck" or anemia of the cobalt-deficient areas of Florida, Vermont and Michigan were once looked upon as contagious or at least mildly infectious diseases only to be catalogued finally and simply as deficiency diseases easily curable by changes of diet and the absorption of certain vitamins and minerals. These facts also have a considerable relation to organic material, for in certain areas where certain necessary minerals *did* and do exist in sufficient quantities, these were simply unavailable to the plants and consequently to animals and people in sufficient amounts because the agriculture was or is so poor and the organic content of the soils so low that the complicated processes, still largely unexplored, by which minerals present in the soil are made available, simply could not take place.

The conversion of many of the so-called "worn-out" soils in the Deep South and the Middle South to a better agriculture, including a greatly increased organic content, could conceivably do much to improve the health, vitality, intelligence and economic prosperity of the people living in those areas. It is not only that the soils in those areas are, in fact, very often depleted and eroded but that the practice of a low-grade agriculture renders many of the minerals actually still present in those soils unavailable to the plants, animals and people living on them. A higher organic content in these soils, with the whole chain of processes accompanying it, which makes for living, productive and fertile soils, could conceivably not only raise standards of health and consequently of resistance to contagious and infectious disease but actually cure deficiency diseases such as pellagra. It should be remembered that a change of diet *alone* may not be fully effective if the elements of that diet are grown upon deficient soils and so are deficient in the vitamins and minerals they are *assumed* to have under average soil conditions.

This has been a long-winded and complicated chapter, yet the implications of the discoveries which are being made every day with regard to minerals, vitamins, resistance, immunity, anti-biotics and other factors closely related to soils and even more closely related to organic materials in soils have scarcely been touched upon. At

[78]

Malabar we have long been observing and making deductions and putting them together into what seems to indicate a reasonable and coherent pattern involving minerals, vitamins, anti-biotics and other factors directly related to soil. All of these factors appear to be a part of a definite code of laws, possibly absolute, by which Nature provides the answer to many of our problems and ills of disease, of nutrition, of economy and of vigor and intelligence. All of these laws point away from the total domination of curative or "patent" medicines and disinfectants in the direction of a philosophy based upon natural law itself.

In the past century we have, from the exaggerated use of commercial fertilizer to the panacea of disinfectants and inoculations against diseases, which subsequently were proven to arise not from infections but from deficiencies, done a great deal of muddling. It is a hopeful fact that today one of the principal forces behind medical research is an impulse in the direction of preventive rather than curative medicine and that more and more of the research is leading directly into the field of agriculture, soils, minerals and the nutrition derived from them. Where in the immediate past most of our research was in the field of inorganic chemistry, we are finding today more and more answers to ills ranging from arthritis to anemia originally in the organic field. More and more we are discovering that many ancient superstitions, from mouldy bread poultices to methods of plowing, are well rooted in scientific fact.

In the past these "superstitions" were practiced because "they worked," just as the first old dame who made digitalis tea out of foxglove found a stimulant for failing hearts which is still in use today. It should never be forgotten that there can be and have been "scientific" superstitions as well as the pragmatic ones of the peasant and the witch doctor. In surgery we had the "appendicitis period" when everyone with a mild stomach ache was operated upon and later the "mastoiditis period" when everyone with an earache was attacked with hammer and chisel and many died who would have lived if they had been let alone to get well or had we known the efficacy of sulfa drugs and the anti-biotics. In a world as complex as this one, today's "truth" may well be tomorrow's "error." The specialist with his limited education has led us down many a blind alley and will continue to do so until we discover that all the elements of our existence fall somehow into a pattern. When all of that pattern is discovered, most of our human ills will disappear or at least be greatly diminished.

[79]

CHAPTER V

The Chicken-Litter Story

One of the basic grievances of this older generation against the younger of today, with its social agitation, its religious heresy, its presumptive individuality, its economic restlessness, is that all this makes it uncomfortable. When you have found growing older to be a process of the reconciliation of the spirit to life, it is decidedly uncomfortable to have some youngster come along and point out the irreconcilable things in the universe. Just as you have made a tacit agreement to call all things non-existent, it is highly discommoding to have somebody shout in strident tones that they are very real and significant.

—RANDOLPH BOURNE, American Essays

V. *The Chicken-Litter Story*

AS HAS been observed several times in this book, it is virtually impossible for any man or any group of men to keep pace with the almost incredible advances in the science or even in the agriculture of our times. It is even more impossible for any man or group of men to fathom the underlying significance or co-ordinate the various elements of our scientific advance into a comprehensive and orderly pattern. In this element of mere co-ordination alone lies the key to even greater discoveries, not alone in agriculture, but in medicine and chemistry as well. The intricate interrelationship and pattern in animals and people of the glandular system and the effects upon it of various minerals and elements have been touched upon in an earlier chapter. And it is highly significant that medicine and in particular the special sciences of endocrinology (the glands) and physiology are turning more and more to agriculture and in particular to soils for the answers to many human ills, some of them hitherto regarded as insolvable. So is chemistry, in its exploration of the organic and inorganic chemical reactions and processes which take place within soils, plants and animals.

In this field the whole process of research which might be called "The Chicken-Litter Story" represents better than any evidence I know the almost incredibly intricate pattern involved in the chemistry of the fungi and the moulds which play so important a role in soils, disease and nutrition. The story is comparatively new, the results having been published less than nine months before the writing of this chapter, and many of its ramifications are still unclear and unformulated although they indicate still further revolutionary developments. Certainly the full significance of the story has by no means penetrated the more murky and solidified academic regions of our agricultural or even our medical education. Yet it is doubtful that any story has been of greater importance to the advance of medicine and agriculture since the revolutionary discoveries of Pasteur. So vast and revolutionary was the work of Pasteur and the men who immediately followed him that for a century most of medicine and veterinary theory, speculation and research in science has gone largely down a narrow, specialized alley in which virtually the whole process was based upon bacteria, disinfections, serums and vaccinations.

At Malabar we would be the last to belittle the value of the discoveries and advances in these fields, yet we do suspect that they do not represent the *only* element in the advance of medical and veterinary science nor the *whole* answer to the ills of the human race and in the field of animal husbandry. In looking to *contrived* vaccines and serums and corrosive disinfectants as the whole and only solution to many ills, both medicine and veterinarian science have overlooked many factors of equal and possibly much greater importance. In the discovery of vitamins, enzymes and the whole field of the miraculous anti-biotics, such as penicillin, streptomycin, etc., immense fields of knowledge have been opened up which in turn have upset and are daily upsetting the whole medical philosophy based upon vaccines, serums and disinfectants alone. The rapidly developing knowledge in these particular fields (vitamins and anti-biotics), together with the increasing knowledge regarding the trace elements and their effects upon plant, animal and human health, has by no means been as yet co-ordinated and, indeed, in many academic circles not even understood (aside from the fact that all too often there is no will to understand them but only to deny them because they are, to the lazy or petrified mind, disturbing).

At Malabar we have had from the beginning an intense interest in the developments of medical science touching upon the relation between soils and disease, and in our own observation and practice we have always allied ourselves with the school of preventive medicine rather than that of what might be called "patent" or "disinfectant" medicine. We have preferred wherever possible to prevent our plants, animals and people from becoming infected by disease rather than to attempt to cure them afterward with vaccinations, serums and disinfectants.

In so far as the Neo-Pasteur-followers have produced well-proven vaccines and serums which provide immunity in normal cases, they have been on the right track and could properly be said to be practicing preventive medicine; but, and this is a very big "but," all too many claims are made in the shadowy realms of certain diseases for vaccines and serums which have not been solidly effective and for disinfectants and frequently for vaccines and serums which often enough create as much as or more harm than good. As one of my scientific medical friends said not long ago,

[84]

The word "virus" has become a wonderful catch-all to explain what we do not understand. If we find ourselves up against a disease which proves to be a medical mystery, we say that it is caused by an unfilterable virus which we cannot see even through a microscope, and then set to work to try to find a "serum" which will cure or prevent the mysterious disorder which we do not understand. A good many times the serums are about as effective as so much distilled water and sometimes they are actually harmful and even poisonous. It is not impossible that many so-called "virus diseases" (which is simply a face-saving name for something which defies for the moment our intelligence and research abilities) may well turn out to be diseases caused by dietary and mineral deficiencies which operate in many intricate and mysterious and complicated ways in certain more or less fixed regions both of anatomy and geography to create the impression that they are infectious diseases.

The cases of cobalt in relation to a special kind of vitality-destroying anemia, the case of iodine and thyroid disorders and the consequent inability of sufferers to absorb calcium and phosphorus in sufficient quantities, the relationship of zinc deficiencies to leukemia, all have been cited earlier and there are many others like them.

All of these disorders were at one time or another—some in the remote past—suspected of being infectious disorders or were attributed to hookworm or heredity or bad diet or other undoubtedly contributing factors, but in each case the basic cause was proven eventually to be deficiencies of certain minerals or elements which, often enough, deranged the normal functioning of other organs apparently only remotely connected with the disease. It is by no means impossible that many other diseases of animals and people, believed to be infectious or even contagious (as many specialists on Bang's disease believe), may be cleared up and in many cases controlled or eliminated once they are treated as deficiency diseases or diseases aggravated and promoted by deficiencies arising from inherited or congenital physiological weaknesses.

It is undoubtedly true that if in the West we had adopted the Chinese method of paying doctors for their services we should have made a much greater progress in the direction of preventive medicine, for the Chinese pay their doctors only while the patient is in good health, and when he becomes ill, the payment ceases until the patient has recovered. The same system would undoubtedly have done much to allay the clamor for socialized medicine. Perhaps the greatest fault of our medical science has been its almost total emphasis upon curing a patient *after* he has become ill rather than upon preventing him

[85]

from becoming ill in the first place. Few diseases could be said to be beneficial. None can be called directly so, and all leave destruction and an increased weakness in a greater or lesser degree in their wake. All of this applies equally to the philosophy and practice of much of our veterinarian science.

At Malabar we have always sought, both in the field of soils and in the health of plants, animals and people, to find out how best to work *with* Nature in the profound belief that Nature was on the whole benevolent and could give us the answers in the fields of health as well as economic prosperity. This is, of course, no more than the philosophy of the great scientist or research expert working in the laboratory. He is seeking to find the answer from the laws of Nature herself rather than to concoct a Swamp Root Elixir or Snake Oil which cures everything from warts to glanders in man, child or beast. Too much of our medical and veterinary research and philosophy during the past century has been confined to the Snake Oil field.

At Malabar we have tried, humbly enough, during the active life of a busy farm to find some of the answers, and to our own satisfaction and belief we have frequently done so or at least have found a hint of the solution or some factor which helped us to tie together several other apparently isolated factors. In the comparatively simple cases of soil erosion, water control and soil improvement, we had great amounts of practical knowledge and advice of immense value from professional sources, and to these we have made our own contributions, some of them wholly new. In the field of the relation of these things to the health of plants, animals and people, there was and still is comparatively little information or knowledge available, and most of our practices and discoveries have grown out of observation of what went on about us and out of a kind of instinctive feeling of what produced the best results. Put quite simply, this instinct meant simply the capacity to imagine one's self as a head of lettuce, a cow, a chicken or a cubic foot of soil and then conceive what would be best for the particular plant, animal or cubic foot of soil, because even cubic feet of soil on a single farm can vary immensely whether, as in our case, it be rich Wooster Silt loam or less rich Muskingum shale or a rich but stiff and stubborn clay or, as in some cases, a cubic foot of almost pure washed glacial gravel.

It was this instinct which led us into many practices which later proved highly profitable, particularly in terms of health and vigor and resistance to disease. It led us to the prohibition of ever putting any animal to live upon concrete. Instead we chose always our own good gravel loam soil which could not be thoroughly disinfected and which, as it turned out, had no real need for disinfection since natural law took care of that. It led us to the pen-stabling of cows where they run loose in big sheds in deep straw and only contact concrete when they stand for a few minutes in the milking-parlor where the concrete can be washed down with water under pressure immediately after milking. It led us into the accumulation of manure covered each day by fresh straw on which our animals lived. It led us into the cafeteria feeding of chickens and the twenty-four-hour-a-day feed program in the cattle barns. It led us into concentration upon the maximum use of organic material for restoring the availability of our natural soil fertility and that of the commercial fertilizer we bought. It led us into many other things—simply this factor of imagining ourselves as cattle, chickens or soil and then working out what would be best for us.

As to the stabling of animals on concrete, the thermometer showed that in zero weather the temperature of the straw under a cow lying on concrete was frequently some 30 to 40 degrees below that of the temperature of the fresh straw on the accumulation of warm manure in the loafing-shed. One had only to imagine one's self a cow and ask how we should like an udder resting in zero weather on cold straw, covering thinly a layer of hard and icy concrete. We asked ourselves how we should like being locked into a stanchion unable to move about or to lie down except in a fixed and cramped position on cold concrete for four to five months of the year. We asked ourselves how, as a cow, we would like being given the exact amount of feed twice a day without any choice as to our own particular appetite or preference. We might, as a cow, like twice as much dry hay and half as much silage as we were arbitrarily fed or vice versa.

There appear on the surface to be certain comic aspects to such a philosophy, but it is a far better and more accurate means of arriving at efficiency and production and health and profits than that of sitting in an office chair working out, remote both from soil or animals, arbitrary percentages of this or that for the assumed good of elements with which there is little or no contact. It is in line with our general policy that any man handling livestock must first love and understand the animals he is handling and possess an imagination which makes him of value to the animal and consequently to the farm as a whole. If he has further technical education so much the better, but no amount of college degrees would bring a man into our employment unless first he had an instinctive love and understanding for livestock. Without those qualities, no matter how much education he possessed,

[87]

he would in the long run be a liability, sometimes in a very damaging way. Indeed I have heard that one of the great professors of mechanical engineering at the Massachusetts Institute of Technology always began his lectures by saying, "Young men, I can teach you facts and you can find them in books. I can make you into mediocre engineers but if you cannot imagine yourself to be a steel girder and imagine how you would feel and what would take place under heavy stress you will never be great engineers."

It is this element of creative imagination which, of course, makes the difference between a scientist and the laboratory researcher who may never rise above the level of a drug clerk.

One of the tragic factors in our American agriculture is the number of farmers who *hate* both their land and their livestock. They are the farmers who are on the land because they never had enough gumption to get off it. They are the farmers who produce very little more than they consume and represent largely a liability rather than an asset to the economy and well-being of the nation. They are the farmers who are kept on the land by agricultural subsidies while they continue to ruin it and to gnaw slowly away at the economic welfare of the rest of the nation. They are the farmers who hate their more prosperous neighbors who farm well and hate the county agent and the soil conservation engineer. The truth is that they are not farmers at all, but merely shiftless citizens or individuals who have never discovered what it is they want to do in life, if indeed they want to do anything at all. They are, by and large, the same group over which the sentimental "liberal" sheds crocodile tears. The tragedy is that there are so many of them, *without* love or instinct for their soil or their animals or, indeed, for anything at all. In them the instinct or genius of which I am writing is almost totally absent and would have no significance.

It was this instinct and the process of imagining one's self to be a chicken which led us into "The Chicken-Litter Story" by the back door.

During the war we raised a good many chickens, both heavies and layers, and we had endless trouble with coccidiosis, with range paralysis, with cannibalism and other ills. Each year for several years we changed hatcheries in the hope of escaping range paralysis, but the incidence of the disease remained about the same. At one time we even resorted to the ridiculous measure, recommended by a poultry "expert," of putting guards on the beaks of every hen to prevent her from attacking and pecking to death other hens. The mere sight of the silly appearance of the flock of birds apparently each equipped with spectacles should have told us that once again this was not an answer at all and cured nothing. It was merely another example in very concrete form of the Swamp-Root-Snake-Oil school of agriculture and animal husbandry. The answer of course was a change of diet which corrected the deficiencies of protein, vitamins and minerals. We *did not*, however, practice the supreme silliness of actually putting spectacles containing colored glass on the hens (as advised by another "expert") on the theory that the color of raw flesh on a victim hen would be disguised and so save her from the attack of her protein-starved and cannibalistic sisters, nor did we paint the windows red as advocated by some of the Snake-Oil school in order to produce the same illusion without the expense of sending the hens to an oculist.

We had been told again and again that hens must not be allowed to run over the same ground year after year and that we must keep them shut up in rigidly disinfected houses to protect them from infectious disease picked up on open ground. Yet all the time, under our very eyes lived a big flock of fighting chickens which inhabited the feeding barns, flourished and brought up their young here and there without any disease whatever. Eventually they only died of old age or from the cocks killing each other off. And we had for years a wild Tom turkey and five Bronze turkey hens and always a certain number of pet chickens running together about the grounds at the Big House. This practice, too, we had been told would be absolutely fatal to the turkeys. Unless the foxes got them, all died eventually of old age. And we had the example of neighboring farm wives who raised poultry for the "egg" money and made a good thing out of it without ever troubling even to clean out the litter and constantly disinfect their chicken houses. Among them there was no range paralysis, very little coccidiosis and no cannibalism whatever. Among their flocks which ran free year after year over the same territory, the incidence of other diseases was almost unknown.

Here was the evidence all about us, yet we were told again and again that chickens hatched and growing up and laying in such a fashion had little chance of survival. Among other things, we were told that chicken litter must be cleaned out of the poultry houses at least twice a year and the houses thoroughly scrubbed down with strong disinfectant, especially before a new lot of pullets were brought in. Religiously we followed this procedure, spending expensive man-hours to do what, as it turned out, was a *harmful* practice in so far as the health and productivity of our poultry was concerned.

Despite the careful observation of all the traditional rules, our record on disease and cannibalism continued to be bad until at last in disgust I ordered a whole new regime. In it there were no patent poultry mashes and no more Snake-Oil palliatives and the hens were relieved of the "silly" false beaks designed to check cannibalism. They were put on the cafeteria system of feeding—whole oats and corn raised on the place, our own good green alfalfa hung up regularly in bunches on strings for the hens to peck at, meat scraps, oyster shells and hoppers of minerals containing some twelve trace elements, including notably cobalt, manganese and copper. Each hen was allowed to balance her own diet according to the needs of her own particular metabolism and physiology. Behind this lay our profound and unshaken belief that any animal knows far better than any college professor or commercial feed expert what it needs and in what quantities, provided a sufficient variety of nutritive elements are placed in front of it for free-choice feeding. This is a principle on which the most successful hog breeders have long operated, and it is spreading rapidly into other fields of animal husbandry.

There were in the background with regard to poultry and hog feeding certain other factors which we together with many other farmers knew although we did not know the reasons why. We knew that fresh cow manure fed to hens shut in a poultry house stopped cannibalism and we knew, as do most farmers, that hogs following cattle in the feeding lots of barns grew rapidly and were rarely affected by any disease whatever. This too was a part of "The Chicken-Litter Story" though at the time we did not suspect it nor did anyone know the reason why these elements in the story were facts.

Once the chickens were put on the cafeteria system outlined above, miraculous things began to happen. Within a few months we had no more range paralysis. Cannibalism and coccidiosis disappeared entirely. The egg production increased about 20 per cent and our feed costs fell more than 50 per cent, even by calculating the costs of raising the oats and corn which the hens consumed.

Naturally we were quite proud of ourselves and attributed the *whole* of the success to the cafeteria feeding. The experiment is described in *Malabar Farm* and the whole credit at that time was given to the cafeteria system. Now we know that, while the changed system of feeding contributed a great deal, the real factor was one which we had overlooked as even the keenest observers frequently do. We overlooked it, too, because the pressure of traditional poultry management rules had told us again and again that the missing factor (which we had overlooked) was very nearly fatal to any poultry enterprise. Actually we were even *ashamed* of the element which, as has since been proven, was the most important in eliminating our troubles.

This factor was simply the old, unchanged chicken litter left in the poultry houses with fresh litter piled on top of it from time to time. By this practice we were violating everything that had been taught about changing the litter and disinfecting each time new pullets came into the poultry house. Like most farmers in wartime, we were short of help and had no time to carry through the prescribed sacred practices and the litter remained in the houses sometimes for more than a year without being changed. New pullets were introduced to the same house after the old accumulated litter had been simply covered with fresh chopped straw and *nothing happened*. More than that, disease virtually disappeared and cannibalism actually vanished altogether.

We were not alone in our experience. Much bigger poultry raisers, harassed by the same shortage of labor, had been following the same "bad" practice of failing to clean, disinfect and provide new, fresh litter at regular intervals and they, too, had much the same results with regard to disease and cannibalism. When they asked the answer, the research was undertaken at the Ohio State Experiment Station at Wooster and there began a whole chain of discoveries to which many individuals and agencies have made contributions and which have led to revolutionary conceptions of health and vigor and disease, not only in animal husbandry and the poultry industry but in human medicine as well.

Any observer, taking the trouble to notice the behavior of hens confined perpetually in poultry houses, has noticed that the hens will scratch and peck at old straw litter long after all lost grain has been consumed. We had observed them doing this even when we used as litter shredded sugar cane in which there never was any grain to begin with. This factor was among the first clues.

Investigation showed that on and in old chicken litter, provided it was not actually wet, there came into being a great variety of fungi and moulds. It was then discovered that these fungi and moulds in the old litter itself produced a kind of high protein feed, by the same general process which makes it possible to produce the equivalent of beefsteak out of brewer's waste or high proteins out of sawdust, as the Swedes and Germans did during the recent war. This element

[91]

was named the "protein factor" and actually provided for the hens a high protein feed which killed the protein hunger that in turn caused hens to attack and peck each other to death. Moreover, the fungi and moulds were also creating anti-biotics of the miraculous family of streptomycin, penicillin, etc., which in turn attacked and destroyed the germs of disease. Finally it was discovered that the principal element in the "protein factor" was a new vitamin called B–12 of the utmost importance to the health, growth and reproductive capacity of all animals and people. In turn this vitamin proved quickly to be a far more effective check to anemia and even pernicious anemia in humans than the old liver extract treatment, which was not effective in many cases for a variety of reasons. The results in the case of pernicious anemia proved to be well-nigh miraculous and as a result of the use of vitamin B–12 there are people walking about today, healthy and vigorous, who would have been dead before now but for the vitamin discovered to be one of the products of old used chicken litter.

But the chain did not end there. It was then discovered that vitamin B–12 cannot be formed in the absence of cobalt, one of the key trace elements which was discovered years ago to be the key to the cure of cattle and people suffering from anemia and the "salt sickness" and "droop neck" in areas of North Florida, Southern Georgia, Vermont and Michigan where there was an almost total deficiency of cobalt or cobalt in available form in the soils.

To carry the chain still further, one of the richest sources of vitamin B–12 is fresh cow manure which, as was known earlier, when it is fed to enclosed chickens stopped cannibalism immediately. It also accounted largely for the health and vigor, rapid growth and resistance to disease in hogs following cattle. The cattle actually created vitamin B–12 within the rumen (or stomach) and in the absence of cobalt could not produce vitamin B–12 either for themselves or in excess for the hogs and chickens to feed upon. (It is worth noting that the poultry of all kinds which at Malabar ran free in the barns, and even chickens and turkeys living together with a notable absence of all disease or a fierce resistance to it, constantly had access to fresh cow manure, as well as the high proteins contained in worms, insects and legumes while running free.)

Presently the Wooster station tried a test experiment in which one hundred hens from the same flock, breed and age, together with cockerels, were divided into two groups and both placed upon a protein-deficient diet. But one group was placed upon fresh new straw in a house newly scrubbed with corrosive disinfectant and the other was placed in an old poultry house on accumulated old litter with fresh litter added on the top. Within two weeks an extraordinary difference between the two flocks developed. The hens on the fresh litter in the disinfected house began to droop, laid fewer eggs and the eggs were infertile. In the other house the hens went merrily on scratching and feeding in the old litter, from which they were getting their "protein factor" and vitamin B–12, and went on producing eggs and the eggs proved perfectly fertile. Needless to say, in the first poultry house the hens were left weakened and ready to become the victims of any disease or to succumb to the deficiency itself or to other deficiency diseases arising from such an utter derangement of their nutrition and metabolism as that caused by the deficiency of protein and in particular of the "protein factor" containing vitamin B–12 which was being constantly manufactured by the fungi and moulds in the old litter. Meanwhile the moulds and fungi were busy creating the antibiotics which in turn constantly attacked the disease germs within the litter itself and very possibly proved a far more effective killer than the corrosive disinfectant used in the first house according to traditional practice.

There is one point not to be overlooked—the fact that in cleaning and disinfecting a poultry house (or any other structure housing livestock) the disinfectant kills not only the malignant disease germs but kills as well the benevolent bacteria and the fungi and moulds which are the worst enemies of the malignant germs and simply gives the disease germs a fresh start when they appear after the artificial disinfectant has faded out. In other words, under the concrete-disinfectant school of operation, the machinery of Nature for handling disease germs and, in the case of the old litter, even protein and vitamin deficiencies, is constantly being deranged and the disinfectant practice is constantly destroying the moulds and fungi designed by natural law to cope with such things, thus simply giving disease germs a fresh start and, in the long run, an advantage.

At Wooster, there have now been twenty generations of chickens raised on old poultry litter with a notable improvement in the record of disease, a total absence of cannibalism and a marked vigor and fertility.

To be sure, there were countless indications in all about us of the pattern which finally came together in "The Chicken-Litter Story." The process is but one example of how more and more by the coordination of information among scientists and research specialists new patterns of the utmost importance to mankind are coming into existence, and the co-ordination and establishment of a "pattern" out of the divided, specialized information is perhaps the element of supreme importance.

We knew the effectiveness of fresh cow manure to the health and well-being of hogs and chickens but we did not know the relationship to a still undiscovered vitamin or the absolute need of cobalt in minute quantities in the creation of this particular vitamin. We did not know that the cow created the vitamin in her own stomach and that when there was a deficiency of minute quantities of cobalt she could not do this and consequently herself suffered, starved and finally died. We knew a generation or two ago that "salt sickness" and "droop neck" in certain localized areas were actually a form of anemia, but the cure for it remained completely hidden until its relationship to cobalt deficiency was discovered a few years ago, and of course without cobalt there could be no vitamin B–12.

We had much evidence at Malabar of the workings of trace elements and the effects of deficiencies as well as, contrariwise, the ways in which certain trace elements and balanced and if possible selective feeding would permit the animal or poultry individual to correct its own diet through instinct, provided the necessary elements were supplied. And we had evidence of the workings of fungi, moulds and benevolent bacteria in their attack on disease germs in the open soil as the organic and barnyard manure content of the soil was sharply increased and brought with it a decline in infectious or contagious disorders. In this case it was a combination of a better-balanced nutrition, including even the trace elements, strengthening the animals and heightening their resistance and eliminating as nearly as possible all deficiencies, while in the soils with their greatly increased organic content, and in the pen stable manure and old chicken litter, the fungi and the moulds were at the same time all warring against malignant disease germs.

I have cited in earlier books the disappearance of foot infections picked up in the original cement-like fields as the organic and moisture content was increased and the fungi and moulds were able to live and grow again and go about their task of attacking malignant organisms. It is undoubtedly true that a process closely similar to that taking place in old chicken litter takes place under the pen-stabling-loafing-shed system of dairy operation where the manure and straw is allowed to accumulate for two or three months with deep fresh straw added daily. Certainly in such a process the fungi, moulds and benevolent bacteria have full opportunity to increase and do their work of destroying malignant germs as rapidly as they come into being. It is also undoubtedly true that in transfering this *cured* manure eventually to the fields we are not only supplying nitrogen and certain minerals to the fields but are actually *inoculating* those fields with all sorts of fungi, moulds and benevolent bacteria which will thrive there *provided* the earth contains a high amount of organic material. These same fungi, moulds and bacteria, together with certain acids created in the "cured" manure, undoubtedly possess the power of breaking down mineral elements in the soil and making them highly available to plants feeding on those soils. This in turn makes for a high mineral content in the forage raised on these soils and in turn builds up the constitutions and resistance to disease of the animals and people feeding from those soils.

Over large agricultural areas in the United States there are undoubtedly great varieties of minerals present which are unavailable to plants and consequently to animals and people because the natural processes of a living soil which includes the fungi, the moulds and the benevolent bacteria simply cannot take place because of the adverse conditions created by a wretched agriculture. These areas are given the name of being poor or "worn-out" areas when in truth they only appear so because the soil is dead—killed by a bad agriculture which ignored the replacement of organic materials either in the form of green manures or of the infinitely more important animal manure.

Certainly under our noses for many years has been the example of a well-operating septic tank. The inflow might be loaded with the germs of typhoid, typhus, cholera or any other viciously malignant diseases, but in a well-operating septic tank the water which emerges is harmless, for through the operation of warrior benevolent bacteria the disease germs are annihilated.

Indeed, the further one delves into this whole fascinating and still mysterious realm, the clearer it becomes that some sort of pattern and law, as exact as the laws of physics or astronomy, is in operation by which Nature herself, if understood and given co-operation, provides the means of health, productivity, abundance and fecundity. It is when these laws and balances are outraged that we arrive at disease, sterility and disaster.

The discoveries of all science are not in essence *discoveries* at all but simply the understanding bit by bit of the laws and balances by which the universe operates. In agriculture and animal husbandry and even to a large extent in medicine and veterinary science, this approach has frequently either been overlooked or the form of the pattern has been distorted by the kind of specialist who attempts to find all the answers in his own narrow field. Those who seek short cuts and panaceas and Swamp-Root-Snake-Oil operations and spectacles for hens are important only because they delay and confuse and sometimes block the process by which the truth can eventually be discovered. It is probably true that the co-ordinating scientist with the mind of the Renaissance and the pen of a Darwin or a Huxley can make the most valuable contributions in extending and enriching the knowledge of man with regard to the universe. It is only such a mind which can, standing upon the firm multi-pillared foundation, erected by the modest and sound specialist, construct the whole of the edifice.

CHAPTER VI

Farming from Three to Twenty Feet Down

The essence of man's situation is slowly becoming obvious. . . . A concept recently expressed, speaks of man as now becoming for the first time a large scale geological force. . . . Through the development of physical sciences, funneled into vast industrial systems, he has created and continues to create new environments, new conditions. These extensions of his mind-fertility and his mind-restlessness are superimposed, like crusts, on the face of the earth, choking his life sources. In this metamorphosis he has almost lost sight of the fact that the living resources of his life are derived from his earth-home and not from his mind-power. With one hand he harnesses great waters, with the other he dries up the water sources. He must change with the changing conditions or perish. He conquers a continent and within a century lays much of it into a barren waste. He must move to find a new and unspoiled land. He must, he must . . . but where? His numbers are increasing; starvation taunts him—even after his wars, too many are left alive. He causes the life-giving soils for his crops to wash into the oceans. He falls back on palliatives and calls upon a host of chemists to invent substitutes for the organized processes of nature. Can they do this? Can his chemists dismiss nature and take over the operation of the earth? He hopes so. Hope turns to conviction—they must or else he perishes. Is he not nature's crowning glory? Can he not turn away from his Creator? Who has a better right? He has seemingly "discovered" the secrets of the universe. What need then to live by its principles!

—FAIRFIELD OSBORN, Our Plundered Planet

VI. *Farming from Three to Twenty Feet Down*

I HAVE quoted in the Foreword an old Chinese saying that "The best fertilizer of any farm is the footsteps of the owner." Put another way, one might say that a farmer can learn as much from his own land as any college of agriculture can teach him—if he keeps his eyes open to what is going on around him in his own fields. It was thus that we discovered at Malabar the full extent of the great fertility of our deep good subsoils and began to feed our cattle on forage from three to twenty feet down and give the old, tired, depleted topsoils a rest.

At Malabar Farm we took over four "worn-out" farms and one fair one. They had reached that point where no tenants would any longer undertake them and where even the neighbors hesitated to rent a few of the fields. Some of the farms were completely abandoned with industrial workers occupying the houses. Yet we knew that, fundamentally, all this was rich land, with soil of glacial origin known as Wooster Silt loam, largely the terminal moraine of the second great glacier deposited on top of a clay and sandstone base. We also knew the history of that soil since the coming of the first white settler.

That first settler found our hills covered by thick and beautiful hardwood forest which in summer achieved a tropical luxuriance and density. For a million years the deep-rooted trees had grown deep down into the deep glacial subsoil, gradually absorbing its rich inorganic mineral fertility and translating it through leaves, twigs and trunks into organic form which was deposited on top of the original subsoils. By the time the first settler arrived he discovered, once the land had been cleared of trees, a rich subsoil filled with organic material to a depth of a foot or more. Its fertility, unlike that of some pine forest areas, was very great, nearly as great as the soils of our rich virgin grasslands.

Once the forest was cleared, the settler raised bumper crops without fertilizer and for three or four generations his successors did well indeed. Then the land began to fail and went downhill in production from then on until the day we took over. During one hundred and thirty years the land was farmed indifferently or poorly, but the most damaging factor was that it was farmed only eight or nine inches deep for one hundred and thirty years, often without any attempt at re-plenishing by any means whatever the mineral and organic content of that eight or nine inches of soil. Moreover, under an almost constant agriculture of open-row crops with soil lying bare the year round the process of leaching or washing out or downward the fertility of this shallow top layer had been devastating. By the time we arrived one of two things had happened: either the remnants of that originally rich topsoil had been completely eroded away, or, where the soil remained, it had become so depleted that wherever a groundhog dug a burrow, the good rich gravel and silt loam he brought from the subsoils to the surface acted like fertilizer on the topsoil.

That is a condition one can witness on farms throughout the United States in any area inhabited by the groundhog tribe, even upon farms which have been managed reasonably well. Keep your eyes open the next time you drive through the country or observe your own fields and very likely you will see the same thing—small spots in fields of hay, wheat or oats where the crop is greener, higher and more productive. Walk over to them and, unless the farmer has spilled some fertilizer, you will find that our friend the groundhog has dug himself a home and spilled subsoil over the topsoil. Our own groundhogs told us a great deal—that the mineral richness of our deep subsoils was very great and that the remaining topsoil was de-pleted to a point where, in the beginning, some fields would not pro-vide yields which merited harvesting.

We had the usual superficial soil tests made and the results were far from encouraging but, as we found out later, they were purely superficial and often misleading. The tests were run to the depth of about eighteen inches and they showed, considering the low yields, a surprising amount of potash and phosphorus but an almost total lack of calcium. What the tests did not tell us was whether the potash and phosphorus were available. With regard to the potash we discov-ered the superficiality and error of the simple soil test within the very first year.

At that time the whole acreage of the farms would not feed ade-quately twenty-five head of cattle, winter or summer, and we were forced to raise Wilson soybeans for hay. We put out a large field with 250 pounds to the acre of 3-12-12 fertilizer. As the summer progressed the field looked yellow and miserable and the beans showed the brown spots of potash deficiency except in irregularly placed large circular areas resembling gigantic polka dots which appeared here and there in the field. Here the hay beans grew rich and rank and were a dark green. All through the summer the odd appearance of the field puzzled us. This time it was not the work of groundhogs. By the end of the summer, after much reflection, I hit upon the reason for the handsome, healthy green polka dots. At some time, certainly generations earlier, perhaps a century, when the forest had been cleared away, the brush and logs had been piled and burned, and where this had occurred there had been created great residues of potash in highly *available* form, so great that they showed up generations later in a field where otherwise the potash had been used up by shallow farming or had become locked up and unavailable. After that experience we increased our potash fertilizer with good and expected results.

But it was not until we got thoroughly into a program of deep-rooted grass and legumes that we began to find some astonishing answers to a great many of our problems ranging all the way from breeding troubles to low yields.

Gradually over a period of years, we moved through the conventional four-year rotation into greater and greater acreages in deep-rooted grasses and legumes. The faster we moved the greater became the fertility of our soil and the more dollars appeared on the black side of the ledger, until today we raise no more corn or row crops of any kind but only a rotation of deep-rooted grasses and legumes into small grains and back again into grasses and legumes.

Before we were able to raise the legumes we had to put on as a "starter" over the whole of the farm 2 tons of ground dolomitic limestone per acre. This quite literally "poisoned" the brome sedge and the poverty grass and permitted us to grow the nitrogen-producing legumes and the better deeper-rooted grasses which in time crowded out the poor land vegetation. At this point, by studying our own land and plants, we made another remarkable discovery.

During the first year of an alfalfa seeding, when the plants rooted little more than eighteen inches, they displayed signs of many deficiencies—notably, manganese, potash and boron. They had spindly stems and the leaves had brown and yellow spots with yellowish bands along the edges, but in the second year when the roots had penetrated three or four feet into the soil, the alfalfa turned rank and lush and every sign of deficiency disappeared. We made another interesting observation—that our best alfalfa, ladino and grasses grew not on the fields where the badly depleted topsoils remained but where there was no topsoil at all, especially on the banks and slopes where the soil content had as high a concentration of gravel as 80 per cent.

One thing was clear—that the shallow and superficial soil test had been fairly accurate about the condition of our topsoils but it told us absolutely nothing about the conditions in our minerally rich subsoils. Friends from the Ohio Experiment Station at Wooster, with whom we work very closely, made deep borings, to a depth of ten to twelve feet, and the results showed what the alfalfa had already told us—that deep down we had a remarkably high rate of mineral fertility and that the subsoil, instead of being sour, had a pH of seven to eight, almost too much lime!

At this point we opened one of the gravelly banks where the alfalfa, ladino and brome grass grew more rankly than on any spot on the farm. We made a sharp deep cut about twenty feet in depth. Not only were we able to use the gravel to advantage on lanes and even in building, but the gravel pit became a revelation and demonstration for the countless farmers who visit Malabar each year.

The cut revealed mostly deposits of gravel, roughly stratified with thin streaks of sand. It also revealed fragments and glaciated bits of stone and geological conglomerates of almost every kind existing from Hudson Bay to our part of Ohio, most of them in some state of crumbling and decay in which their mineral content became available to plants growing high above them on the surface. There was in particular a very high percentage of limestone in small pieces, ranging from soft chalk through dolomitic limestone, which carries a good many trace elements as well as pure calcium.

In that cross section we found some other fascinating and important evidence—the brome grass in cross section showed a fan-like distribution of fine hair-like roots penetrating to a depth between two to three feet. Orchard grass had an even deeper penetration. The alfalfa roots (on the three-year-old plants) penetrated to an average depth of fifteen feet and the longest measured root had gone down to a depth nearly twenty feet! We could now see where the plants were getting all the mineral nutrition they could utilize and why the second-year and older plants showed no signs whatever of mineral deficiencies. They were getting all they wanted and more.

The cut also illustrated a factor which many farmers tend to forget—that the roots of deep-rooted plants produce great quantities of organic material which later becomes translated into humus. When plowing under green manure, the average farmer is likely to think that he is plowing under only what he sees above ground. The roots have a great importance both as humus and for their high mineral content which is in organic form and therefore highly available to the crop which follows when a deep-rooted sod is plowed under. The cross section also illustrated the factor of good drainage essential to the growth and development of deep-rooted legumes.

It was apparent that, in the program of deep-rooted legumes and grasses, we were, after the first year of a seeding, scarcely using the long overworked, depleted and leached-out topsoil at all but were drawing upon the fertility of the deep subsoils which had not been utilized since the forest had been cut off the land more than a hundred years earlier. After the first year of a seeding the roots by-passed the topsoil and went deep, finding there not only all the mineral fertility the plants needed but a fertility which was in good balance. Under the deep-rooted legume and grass program we were, after the first year, scarcely using the topsoil of our farm at all save in growing ladino clover and our native bluegrass and white clover.

But the story did not end there. The alfalfa, rooting deep in the subsoils, was found, when analyzed by the Batelle Metallurgical Institute at Columbus, Ohio, to contain a mineral content of 6.5, considerably higher than the average mineral content of alfalfas grown in the United States. This amount more than supplied the mineral needs of the cattle to which it was fed and the residue was passed on in the form of barnyard manure which went back onto the depleted, leached and worn-out shallow surface of the land, replenishing its fertility exactly as the original forests had done in building up the originally rich topsoils over a period of thousands of years.

Moreover, each time we plowed in one of the deep-rooted heavy sods, the same process was taking place as the minerals, now in organic form and therefore highly available, drawn from deep down were mixed with the worn-out topsoil. This factor accounted for the astonishing jumps in yield of oats and wheat and, in the beginning, even of corn, which occurred when a manured, deep-rooted sod field was plowed and new crop seeded. Actually we were and are farming those Wooster Silt loam soils from three to twenty feet deep after more than a hundred years of farming them only eight to nine inches deep. And in the process we are undoubtedly recovering some of the fertilizer used by our predecessors on the same land which, in an area of abundant rainfall, had leached down through the loose gravel loam into the deeper strata of the soil.

Other effects showed up presently in the livestock themselves. In the beginning while the topsoils were still leached and worn out, the rate of disease and infections ranging from sore feet to mastitis among livestock was high and even discouraging, despite the fact that, knowing the deficiencies of our soils, we kept constantly before them in a compartment next to the salt box a mineral mixture containing twenty-two trace elements. It is notable that in the beginning, while the farm was still producing low-quality forage, deficient by its own evidence in many elements, the animals consumed very nearly as much of the mineral mixture as of salt. Once the program of deep-rooted grasses and legumes got under way, their consumption of these minerals declined rapidly until today weeks go by when the chaff on the mineral boxes goes wholly undisturbed by the visit of a single animal.

The obvious deduction is that, once we tapped the deep-down fertility, the animals began to get an abundance and more than a sufficiency of minerals of all kinds and no longer sought them elsewhere. They were getting them as they should, naturally, out of the pasture and forage which they consumed rather than artificially in the form of various chemical salts or, even worse, in the form of pills, capsules and injections.

Obviously, the same things would occur to the people living off the vegetables grown on the same land, although the human instinct has long ago lost that force and direction which drives a sow to chew up cement in order to get enough calcium for her pregnancy or a cow to develop a depraved appetite for old bones because the pasture and forage she consumes is deficient in phosphorus and calcium.

The old infections and diseases which plagued our livestock in the beginning have virtually disappeared from the farm so that we have had no visit from a veterinarian in more than three years except for tests or accidents.

Perhaps the most interesting factor is that it was the plants themselves which first revealed to us by the spots and discolorations on their leaves the deficiencies of the top few inches of worn-out soil, deficiencies later confirmed by the sickness and infections of the livestock themselves. After two or three applications of manure and two or three plowings-in of deep-rooted grass and legume sods, the spots and discolorations on the alfalfa leaves during the first year when the roots only penetrate to a depth of eighteen inches have also begun to disappear as the available mineral fertility is restored from deep down.

Another factor which we learned simply through observation and which has been described in an earlier chapter was the gradual disappearance of attack from leafhopper and spittle bug.

All of these facts were discovered almost wholly by observation, close and thoughtful, of our own land which *taught us* how best to treat that land. By the employment of the sweet clovers, annual sweet clover or Hubam which in one season will root down to five or six feet, biennial clover which will root down eight to ten feet in two years, alfalfa and deep-rooted grasses such as brome, orchard grass and alta fescue, we have been able to tap and utilize the almost inexhaustible fertility of our deeper soils. In this process, including as well limestone and a reasonable amount of chemical fertilizer as "starters," we have raised yields of crops on the once abandoned "worn-out" farms as much as ten times in a fewer number of years. Perhaps no area on the whole farm displayed so immediate a reaction as the Bailey hills. I have recorded the story in an earlier chapter but repeat it here since it is so important in the restoration of what was regarded as "hopeless" land.

On the hills exactly opposite the Big House at Malabar lies a 100-acre tract, fairly rough and broken, which when we acquired the land would not adequately pasture five cows during the summer season. It was covered with brome sedge, poverty grass, wire grass, sorrel, briars and other poor land plants. Virtually no topsoil remained and the land was sour. We tried several methods to bring about the restoration to fertility of this land, among them using oats or wheat as a cover crop. Even after being limed, the hills scarcely gave us our seed back, and the seedings of grass and legumes were poor probably from the lack of moisture during the hot dry months on hilly land, totally devoid of organic material, which caught all the sun and wind. Again we turned to Nature herself for the answer and tried seeding the newly limed land by the trash mulch method, tearing up the old weeds, grasses and briars not deeper than three or four inches, mixing the ground limestone and rubbish in the top layer of soil. Then with the addition of 300 pounds of 3-12-12 as a "starter," we seeded in 10 pounds of alfalfa, 5 of brome grass, 3 of orchard grass and 1 of ladino clover.

In the trash-mulched surface there was literally *no* erosion even on the steepest slopes, and the seeds, falling into the mixture of soil and organic trash which retained the moisture, gave us very nearly 100 per cent germination. Every weed seed also germinated, but we took advantage of the weeds by mowing them down before they went to seed, twice during the first season, and leaving them as an added mulch to keep the ground cool and preserve moisture during the hot dry months. Fourteen months later we had a flourishing crop, very nearly weedless, of alfalfa, brome grass, orchard and ladino clover. In the second year we took from the hills a good cutting of grass silage, a second cutting for hay and had forty days of good autumn pasture.

Especially notable was the fact that the alfalfa on these hills showed during the first season signs of fewer deficiencies than on the fields where the old worn-out topsoil remained in considerable quantities. It further verified our suspicion that we could make greater progress in restoring our land by working directly with the good subsoil itself than by trying to build back the worn-out topsoil wherever it remained. On the subsoil it was possible quickly to get good crop yields simply by "pumping in" great quantites of organic material to hold moisture and restart the whole process by which organic material and moisture make the natural mineral fertility of the soil available to plants and consequently to animals and people. As one of the boys put it, "I guess we'd have made greater progress by scraping off the old worn-out topsoil that we inherited and, like Nature, beginning from scratch."

One notable happening was the fact that, on those once barren hills, the land, which would not give us our seed of oats or wheat back in the beginning, would, after three or four years of deep-rooted grass and legumes, give us, when plowed up after that period, yields of 35 bushels to the acre upward of both grains. It was clear that the deep-rooted grasses and legumes had done their work both by providing quantities of organic material and by pumping up the deep fertility into the top level of the once bare, weed-grown, leached-out hills, and all this in addition to the fact that there was no longer any erosion or water loss on those hills.

All of these observations and practical experience are recorded primarily as *observations*. At Malabar we do not pretend to know all the reasons and even today research itself has not provided all the answers. We do know that these things happened and that the listed methods worked and that we are now growing, on land which ten years ago or less was much of it "worn-out" and abandoned land which would no longer pay taxes and interest, crops which in yield total on some of the land two or three times the average yields of the same crops in the rich state of Ohio.

We also know that the expenditures involved were 4 tons of lime to the acre and never, except in the case of trash mulch seedings,

[106]

more than 200 pounds of commercial fertilizer to the acre and sometimes less. It is clear and reasonable, I think, to assume that the tremendous gains in yields could not be attributed to commercial fertilizer alone since we used less than many farmers use per acre and per crop, yet the fertility leapt upward and production in both quantity and quality rose from the lowest possible level on the type of soil we took over to a highly profitable level. It came from down deep, and the gravel pit, with its tremendous accumulation of rocks and minerals of all kinds, showed us where it was coming from.

The observations and practical results have, I confess, led us into many speculations, some of them perhaps a little fantastic. One of them is the speculation that by farming from three to twenty feet down with deep-rooted grasses and legumes we have tapped a mineral fertility which, like that which grew the original magnificent hardwood forests of Ohio, may prove inexhaustible. Certainly as the gravel pit showed us, the minerals are there in the form of rocks, gravel and pulverized stone. The question is whether they break down into availability rapidly enough to supply the full needs of the plants growing far above on the surface. We do not yet know the answer and possibly will not have it for many years to come.

We do have one seeding of alfalfa which, along with brome grass, has gone without barnyard manure or fertilizer for a period of nine years. It is still producing a good crop of hay each year although the increases in nitrogen produced underground by the alfalfa have so stimulated the brome grass that it has tended to crowd out the alfalfa. We are led into the supposition that a top dressing of barnyard manure once every three years would keep that particular test strip in maximum production almost indefinitely. Where does the mineral fertility come from?—From three to twenty feet down. We still make top dressings every three years of commercial fertilizer or barnyard manure on our other heavy grass seedings and give especial attention to the shallow-rooted ladino and native bluegrass pasture since they have not the capacity to reach down deep and find the fertility they need. We do not expect them to "take care of themselves" to the same degree as the deeper-rooted plants.

The history of this particular test strip is especially interesting since the original seeding was made upon a long-abandoned field grown up with broom sedge, poverty grass, goldenrod and other low-quality weeds. It was limed, fertilized and roughly fitted so that actually the surface of the strip was simply a heavy trash mulch about eight to nine inches in depth. Into this we sowed a nurse crop of oats and the scrapings of the seed bins which included red and mammoth and alsike clovers, timothy, alfalfa and brome grass.

The germination was excellent but we did not get back the seed on the oats crop, which was simply left standing. In the second year we took perhaps a ton of mixed hay off the field and by the third year much of the shallow-rooted clover and timothy had disappeared, leaving a better than average stand of deep-rooted alfalfa and brome grass. From then on the shallower-rooted plants went backward and eventually disappeared save for some of the timothy which hung on somehow and staged a comeback once the alfalfa roots began producing great quantities of nitrogen. It is notable, however, that while all the leaves and stems of the timothy were fairly vigorous, the heads were short and stunted, indicating the increasing presence of nitrogen but a shortage of most other elements. The deeper-rooted alfalfa and brome grass continued to flourish, and even though they were growing *through* a badly depleted topsoil, the deficiency signs on their leaves gradually disappeared as their roots penetrated deeper and deeper into the subsoil where the supply of minerals was and is still abundant. In other words, the brome grass and alfalfa have simply by-passed the top exhausted layer of soil which would not maintain shallow-rooted grasses and legumes.

After nine years, without fertilizer or manure, the test strip is still producing about twice the average hay yields of the state of Ohio. This would tend to indicate that deep-rooted grasses and legumes in our Malabar soils could go on producing indefinitely, *provided* the subsoil minerals, mostly in the form of gravel, break down rapidly enough to supply the full needs of the plants from year to year. This we do not know and will probably not know definitely for another ten years.

And above all we do not pretend that every subsoil would give the same results as we have achieved on our Wooster Silt loam with its high mixed mineral and gravel content. Probably very few and only good deep subsoils would react in the same fashion, but there are in the United States millions of acres of such soils and millions of acres of them today are lying idle as abandoned "worn-out" farms, not because they are really "worn-out" but because they have been farmed only eight or nine inches deep for generations and because the residue of minerals which has not been depleted has simply become unavailable to plants because of poor farming and the lack of lime, organic material and commercial fertilizer "starters."

As great areas of the Deep South have discovered, these soils can be restored and brought back to high fertility very rapidly at a comparatively low economic cost. In a word, all we have done on the fields at Malabar is to restore the same processes by which the great hardwood forest flourished for thousands of years and constantly built up more and more rich topsoil. Today we are doing much the same thing by farming from three to twenty feet deep, pumping up new fertilizer in organic form from sources deep down which may be virtually inexhaustible.

CHAPTER VII

Water and the Farm

Sometimes it looks to me as if we'd got to go all the way back to the Indians and begin all over again to undo all the damage our grandpappies have done to this poor country.

—Anonymous Member of a Veterans'
Agricultural Training Class

VII. *Water and the Farm*

WHEN first we took over the fields at Malabar as much as 80 per cent or more of the water which fell during a sudden cloudburst or even during a long heavy rain ran off the farm within twenty-four hours. This was a process not confined to our own acres but in some degree or other it occurred over the whole state of Ohio.

During the century or more since the magnificent hardwood forest which once covered the whole of the state was cut away and the soil put into cultivation, millions of acres were left stripped and bare during the long winter months and, under row-crop agriculture, were susceptible to heavy water loss even during the thundershowers and cloudbursts of the long hot summers. Then, gradually, under an agriculture which failed to take account of the values of ground cover and organic material, the soil of Ohio (save in the hands of the occasional farmer) became depleted of organic material and presently began to assume the texture of cement. As a result two disastrous developments occurred: (1) in the rough and hilly country the soil, no longer loose and porous and sponge-like, refused any longer to absorb the rain as it fell and each year more and more water ran off the fields instead of sinking into the earth to make available the natural fertility and the purchased commercial fertilizer and to feed the deep reservoirs of springs which began presently to dry up while the level of water in wells began to lower with the passing of each year; (2) in the flat country where violent run-off and the consequent loss of topsoils was less of a problem (although by no means a non-existent one) the steadily increasing cement-like condition of the soil prevented the penetration of rainfall to the deeper levels and sealed up expensive drainage tiling so that it no longer served its purpose. Water, on fields which had once drained well without tiling, now stood on the surface to be evaporated at last by wind and sun without ever benefiting the crops at all but often enough creating actual damage by drowning them out.

Both erosion and defective drainage came about very largely through the practice of an agriculture which year after year produced cash crops and neglected the replenishment of organic material

[113]

burned up and destroyed by incessant plowing and fitting in a cash-crop agriculture. Moreover, the incessant plowing and fitting for cash grain crops of these increasingly badly drained, flat-country soils not only destroyed all reserves of organic materials but presently developed a hard-pan like cement about nine to ten inches below the level of the surface and this in turn further blocked drainage, leaving the water in the top level to choke off further agriculture and to remain thus in actual pools until it was at last evaporated by the action of wind and sun. Only in the case of a few good farmers and farmers given over to livestock enterprises, or among the Amish and Mennonite and Dunkard colonies where organics are worshiped as God, did these conditions fail to arrive in some degree or other.

Today there are a million or more acres of eroded and depleted but potentially rich hilly farm land of glacial origin lying abandoned and sometimes even tax delinquent in Northeastern Ohio. In Western and Northwestern Ohio there is another million or more acres of rich land abandoned and useless because the soil structure of fields, which once raised bumper crops with little or no artificial drainage, has so deteriorated that the fields are no longer tillable because they cannot be drained even with tiling placed at sixteen feet apart.

In effect, the process over a great part of the state was to divert the rainfall from sinking into the soils as it had once done in the days of the great hardwood forest and turn it off the land in flood proportions down the Mississippi into the Gulf of Mexico or into the Great Lakes watershed leading into the Atlantic. Even the woodlots, which are virtually all that remain of the once vast forests of Ohio's surface, were heavily pastured by sheep and cattle, which found in them little nutrition, and in their efforts to find enough to eat, destroyed all ferns, seedlings and ground cover. I have seen great gullies in such pastured woodlots, sometimes created by the rush of water from the cement-like soils and bare surfaces of adjoining fields. In the beginning we had many such gullies through the forest at Malabar.

As a result of these processes, by which the water was diverted from the underground reservoirs, and by the constantly increasing use of underground reservoirs of water in industry, for metropolitan water supplies and in air-cooling systems, springs have dried up over large areas, the water level in wells is constantly sinking and the water table as a whole throughout the state has dropped an average of more than forty-five feet in a generation. Forty-five feet is the

Spring carpet in the jungle at Malabar. The jungle is a wild, swampy area given over to wild life and native flowers, shrubs and trees. (*Joe Munroe*)

Baron Ochs von Malabar, commonly known as "Stinker." (Joe Munroe, courtesy The Farm Quarterly)

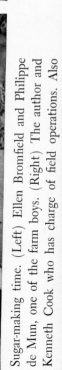

Sugar-making time. (Left) Ellen Bromfield and Philippe de Mun, one of the farm boys. (Right) The author and Kenneth Cook who has charge of field operations. Also

David Rimmer among hardwood forest seedlings which are developing into a new forest since cattle were fenced out of woodlots. Malabar is one of the "tree farms" set up by the Ohio Division of Forestry. (*Joe Munroe*)

(Above) The Seaman field tiller in operation on a heavy sod. (Below) The soil after the tiller has passed over it showing loose mixture of soil with organic material. (*Joe Munroe*)

A two-year-old plant of ladino clover, illustrating large amount of high protein forage resulting from a single seed of this greatest pasture-producing member of the legume family. (*Joe Munroe, courtesy* The Farm Quarterly)

One of the drainage tanks at Malabar, adapted from good European practice, which reclaimed a large acreage of rich, but useless, wet land. (Joe Munroe)

Some of the crew. From left to right, Kenneth Cook, his son and assistant Jim, a younger son George, Bob Huge, and Eldred Bryner. (*Babson Brothers Co.*)

(Left) The finish of a cattle drive at Malabar when all visitors, guests and dogs are put to work together. (*Ferguson*)

Filling the silos with grass and legume silage for the winter. (*Joe Munroe*)

The Ferguson tiller, Jim Cook operating, with the Bailey Pond and buildings in background.

(Above) The Mason place in spring. The desolation which follows bad agriculture on good land. (Joe Munroe, courtesy The Farm Quarterly) (Below) Malabar Farm as it is today reclaimed. It lies over the hill in the next valley from the Mason place. (Joe Munroe)

(Above) The Ferguson hay rake, a genuinely modern piece of farm mechanization. (*Ferguson*) (Below) The field and pasture irrigation system creating a small Niagara with author and Folly in foreground. (*Samuel A. Musgrave, courtesy Gorman-Rupp Co.*)

(Above) Jim Cook adjusting rubber silage cap to seal silo against spoilage. (*Joe Munroe*) (Below) Digging a farm pond with modern farm equipment; one of five at Malabar used for swimming, skating, fishing, irrigation and water for livestock. (*Ferguson*)

Above ground, the heavy production of high protein, high mineral legume and grass ensilage and pasture. The mixture is alfalfa, brome grass and ladino. (*Joe Munroe, courtesy* The Farm Quarterly)

Below ground, the author shows the root of a three-year-old alfalfa plant reaching fifteen feet and more into the virgin, mineral wealth of the gravelly subsoil. (*J. C. Allen and Son*)

Five-year-old hedge of Rosa Multiflora. (Left) Close-up of hedge showing tangle of thorny growth which is both paradise and protection for game and song birds. Top left can be seen thousands of clusters of rose pips which provide food during winter for all sorts of birds and game. (Above) Farm lambs, in a field bordered by Rosa Multiflora, are entirely pasture and forage fed, never tasting grain or supplements of any kind, yet they bring top market prices. As can be seen the hedge could turn back not only sheep but cattle and even hogs. (*Joe Munroe*)

The Graham-Hoehme "plow," Jim Cook up. This is the implement which has largely replaced the mouldboard or turning plow and even the one-way disk, over vast areas, with the result that erosion and dust storms have been checked, rainfall and moisture conserved and bumper crops of wheat produced in the great plains year after year since 1938. (Joe Munroe)

Al Bryner, herdsman, who loves and understands his animals, with Inez, a first calf Holstein heifer from the Malabar herd. She was fed only on hay, silage and pasture until coming into the milk parlor. (*Joe Munroe*)

Pasture mowing on Mount Jeez. The constant mowing of rotated pasture at Malabar has increased carrying capacity as well as quality of feed values by 30 per cent. (*Ferguson*)

Contour strips on the Anson place. Once such strips were necessary to check erosion even on fields planted to small grains. Today, with the increase of organic materials and the emphasis on grass farming, this is no longer necessary. Soil erosion is the result of bad farming. (Joe Munroe, courtesy The Farm Quarterly)

The author and Bob Huge, who played so large a part in the restoration of
the acreage at Malabar and is now Agricultural Director of the Samuel Noble
Agricultural Foundation at Ardmore, Oklahoma, and partner and Director
of Malabar operation at Wichita Falls, Texas. "Stinker" is the third character. (*Joe Munroe*)

(Above) These modern dwellings at Malabar Farms, Wichita Falls, Texas, built for the climate at a cost 30 per cent less than the conventional house, are of pink hollow tile, masonite partitions, aluminum roofing and poured concrete base. No painting, no rotting or rust, no termites. (Below) The modern cattle shelters on the same farm are of pole and aluminum construction. (Ferguson)

The pond at the Bailey place. Both ponds are stocked with bass, sunfish, bluegills and rainbow trout. (*Ferguson*)

The pond at the big house, showing Babylonica willows, grown in three years from switches stuck in the earth. These willows protect dams and prevent stream bank erosion. (*Joe Munroe*)

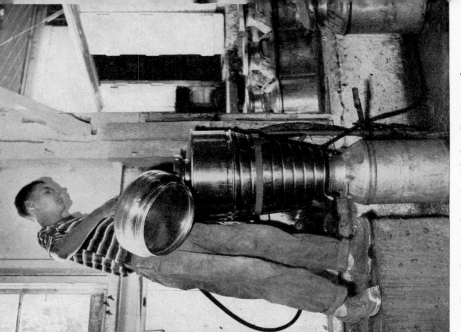

Simon Bonnier from Stockholm, Sweden, who came to Malabar to learn dairying and grass farming. (Joe Munroe)

David Rimmer who came to Malabar from Chester, England, to take charge of plant nutrition and soil experiment. (Joe Munroe)

The last decaying vestiges of the first farmer-trapper's cabin on the Ferguson place—a symbol of the old, careless, destructive agriculture rapidly being liquidated by the force of economics, regardless of subsidies, parities or price supports. (Joe Munroe, courtesy The Farm Quarterly)

A typical "worn out" meadow in the Malabar country. Not only is the grass weedy and thin, but the nutritional values are insignificant. A cow fed on this might still be starving to death. Most of Malabar originally looked like this barren, sickly field. (*Joe Munroe*)

A rich mixture of alfalfa, brome and ladino directly across the road from the sickly field on the opposite page. The figures lost in rich forage are dogs. (*Joe Munroe*)

Heavy strips of wheat and grass and legume mixtures. The lodged wheat was caused by the amount of free nitrogen produced by legumes—such as those through which the two small boys are struggling—without use of commercial nitrogen fertilizer. (*Joe Munroe, courtesy The Farm Quarterly*)

average and in many areas it has dropped a hundred feet or more. In some regions, once abundantly supplied with water, farmers are forced to haul water to their livestock during many months of the year, and some towns are unable to bring in new water-using industries or even to increase their populations to any great extent because of the shortage of water. This is in the state of Ohio with about forty inches a year of well-distributed rainfall.

Methods have already been undertaken to replenish the underground reservoirs by directly pumping the water from surface streams into the known underground cavities. This, of course, is only a makeshift operation and no final solution to a condition which under a continued poor land use can conceivably reduce certain areas, in summer at least, to a semi-desert status. While the increasing use of water for industrial and metropolitan use has aggravated the situation, the greater part of the shortage can be attributed to a poor cash-crop agriculture practiced widely for more than a century. In effect the rain which once fell on heavily forested land and remained where it fell and was absorbed, finding its way at last into underground reservoirs, has been increasingly diverted off the surface of the whole state and down the rivers for more than a century.

At the same time flood conditions downstream on the rivers of the state and in the Ohio and the Mississippi have been immensely aggravated. While in the days of the Great Forest floods actually occurred, they never attained the violence or the volume of those experienced in Ohio during the past two or three generations. There is, indeed, plenty of actual living evidence of this increasing violence. In the early records of Blennerhasset Island in the Ohio River where Harmon Blennerhasset constructed for himself a great and romantic manor house at the end of the eighteenth century, there are no records of floods covering the island, yet today scarcely a spring passes when the waters of the Ohio do not threaten or actually cover the island, sometimes to a considerable depth. And in all the river towns, both on the Ohio and the Muskingum, there exist two towns, the old one located at the level of the river on the bank itself and the second on much higher land where the inhabitants built houses not through choice but to escape from the increasingly violent floods.

Malabar Farm occupies several hundred acres of land largely spread across the saddle of the watershed at the end of Pleasant Valley where the valley of Switzer's Creek melts into the valley of the Clear Fork and the big lake which is a part of the Muskingum Flood Control District and Conservancy. Its acres include nearly the whole of the watershed on two sides of the Valley while the waters of Switzer's Creek descend from an upstream watershed about nine miles long and three or four miles wide covered by farms and wood-lots. It is rough and hilly country with numerous springs and brooks and there are at least three types of soil. At the end of Pleasant Valley and especially on the edge of the lake there is some fairly flat land of glacial gravel character.

The situation of the Malabar holdings gave us an opportunity to observe closely the behavior of the soil and water and forests of the area, and after nearly eleven years their behavior under changed management has brought about results which are little short of miraculous and can no longer be doubted. Our good neighbors, following recommendations of the county agents and soil conservation technicians and obeying the dictates of their own intelligence, have contributed much to the changed conditions. Owing to the situation and extent of Malabar Farm, there can be no doubt as to the reasons for the changes. It might be described as a gigantic test plot in soil and water management.

A period of nearly eleven years has passed since the first measures to control water and soil erosion were attempted and, although the benefits continue each year to increase, there is no longer any doubt of the sources from which they spring. These immense benefits, measurable in health, in morale, in dignity and in dollars and cents, have arrived not only on our own land but on the land of many of our good neighbors simply by a change and improvement in land use.

When we came into the Valley not more than a half-dozen farms in the entire watershed could properly be said to have made any attempt at modern agriculture or proper land use. Today not more than one in ten have failed to change over and to be operated in some degree with proper regard to soil erosion, water loss and higher crop yields.

The change in the character of the small creek which drains the Valley has been especially noticeable. During the early years the whole of the bluegrass bottom at Malabar was inundated every time there was a considerable thundershower or cloudburst. The stream became wildly turbulent and filled with silt and each year deposited on our lower fields from an inch to an inch and a half of topsoil, lime and fertilizer belonging to the farmers upstream. The banks of fields were washed out and the loss in the whole valley, coming from eroded stream banks alone, in actual dollars and cents amounted to many thousands of dollars annually.

This drain, through the cutting away of good bottomland soils by a flooding and wandering stream, had been going on for nearly twenty years since some ill-guided farmers and county commissioners conceived an idea (from where or upon what authority I do not know) that the original course of the stream ought to be straightened. It was, perhaps, a part of the costly and whimsical epoch in American agriculture, some years past, when it became the fashion to ditch and tile everything in sight, to straighten all streams and in general to make a general attack upon the undefeatable laws of Nature. The mania for ditching, tiling and straightening has cost American hunters and fishermen untold amounts of sport and the farmers and the nation millions of dollars worth of good soil and productivity. It also aggravated enormously the problem of failing water supplies.

In Pleasant Valley the original banks of the pretty, clear, little stream, covered with willows and alders and other vegetation, were ruthlessly ripped up, and a drag-line operation created a straight and characterless canal down the dead center of the Valley. Together with the willows, the alders and the underwater vegetation, went the old fishing and swimming holes which I had known as a boy, and with all of these things went nearly the whole of the big fish population.

Nature, however, only laughed at the foolishness of the men responsible for the absurd and costly operation. Within six months the stream began wandering about once more, finding its own level, cutting in on the one side to rich bottomland fields and carrying them downstream, and on the other piling up silt to block the outlets of the tiling on which the farmers had spent so much money in order to drain those fields. Every year the stream altered its course, eating away more and more of the best land, and as row-crop cultivation without strips or contours increased, the amount of run-off water and its destructive violence increased in exact ratio. There were no longer any permanent fishing or swimming holes, and the increasing silt cut from the stream banks and fields drove out virtually all game fish which were replaced by carp and bullheads and suckers. The project probably cost in the neighborhood of $25,000 upward and the farmers in the succeeding years lost twice that amount of good soil from their bottomlands. Why? Nobody knows.

For seven of the ten years or more at Malabar we struggled to find the means of bringing back stability to those banks, of once more covering them with vegetation so that the stream would stop its expensive wandering and build up once more the fishing and swimming holes I had known as a boy. We tried everything recommended by the Department of Agriculture, by the state colleges and by the Soil Conservation Service. None of the recommendations succeeded wholly, although in some spots we did arrest the costly wandering of the stream with sod or willow plantations. And then by accident we discovered a method which proved absolutely effective and by which we have succeeded in once more keeping the stream in its course and restoring the old deep holes between the long gravel-studded riffles. Also it was a method far easier and cheaper to carry out than any of those suggested.

In the first year we came to Malabar Dan Quinn, the nurseryman, brought in two specimens of a semi-weeping, ornamental willow called Babylonica. Within three years we marked the tree as being of as much potential practical use as of actual ornament. The Babylonica never acquired scale nor harbored voracious insects, and the benevolent ladybugs, one of man's best friends, hatched out their eggs and fed their young upon it. When a sleet storm hit the Babylonica, it yielded and bent to the ground, but once the ice had melted it sprang up again into an upright position without the loss of a single twig. And it grew at a prodigious rate. A switch stuck into the earth in the spring produced a thirty-foot tree within three to four years. It never littered the earth as most willows do with dead twigs and branches, and its semi-weeping style of growth brought into the landscape the kind of misty beauty that is in Corot's landscapes.

Each spring for the first three or four years, I seldom went walking over the farm without carrying under my arm a bunch of willow switches, and whenever I found a likely spot for a willow I thrust the switch into the ground. Wherever I stuck a switch there are today trees from fifty feet upward in height. Many of the cuttings I stuck along the banks of the wandering stream, but most of them were washed out before they got a real foothold and sometimes the cattle destroyed them.

The first use we found for them was on the banks of the farm ponds which we built. In the first years, before the whole farm was converted to good land use, there were sudden torrents after heavy thunderstorms which threatened to wash out the dams that held the farm ponds. The trunk roots of the Babylonica willow, we discovered, are much smaller than those of the average willow. Most of their roots made a fine network which bound together the soil of the dams so firmly that if a torrent came, the water ran over the dam without eating into it or causing the slightest damage. This procedure was contrary to all recommendations because the authorities claimed that the big roots, especially when a tree died, permitted the water to seep through the dam and thus started leaks which later became disastrous. This was not true, at least of the Babylonica, perhaps for two reasons—that the roots were so fine and because the Babylonica does not die and, unless cut below the surface of the soil, is virtually indestructible.

After a few years the rapid growth of the willows on the dams developed trees of considerable size which (a) sometimes evaporated too much water, and (b) might blow over causing destruction to the dam itself. So we adapted a policy of cutting them back to stumps every year as the growers of basket willows do. To dispose of the cut branches and trunks David Rimmer, who came from England to run the gardens, threw them along the banks of the wandering creek just above the level of the water; and then something wonderful happened.

The cuttings were thus disposed of in February and early March when the farm work was light, and we discovered that, as spring came along, all of the cuttings carelessly disposed of along the banks began to root and grow, anchoring the cut brush into the bank just at the normal water level. Even bare heavy trunks struck root along the whole of their length and began sending out fresh green shoots here and there along the whole of their trunks. By the end of June the brush had all rooted and the banks were anchored for the summer, and from here on out, Niagara Falls could scarcely dislodge those willows. Where the willow mats advocated by the authorities were swept out again and again, the carelessly cast away willow branches took root and held.

Since that first accident, we have improved the technique. We discovered that when the branches and twigs rather than the butts were placed upstream they tended to collect rubbish and silt brought down by flood waters and that this only anchored them the more firmly to the banks and encouraged the growth by covering the branches and trunks with silt and rubbish which maintained the moisture. Moreover, where the creek took a sharp turn and there was danger of the branches being washed out by a sudden freshet before they were sufficiently rooted, we anchored them by the simple process of driving in a few steel fence posts. After the first summer there was no more danger of their being washed out. Within another year or two we shall have back the old and permanent creek banks covered with vegetation and we shall have back again the old, deep permanent fishing and swimming holes which the foolish creek straighteners had destroyed. We shall lose no more of our good bottom soil, and instead of bare, ugly, washed-out banks, we are getting back the old line of willows which once gave the Valley a singular dream-like beauty.

The process would possibly work with any willow, but with the fast-growing, fast-rooting Babylonica it is certain. We have distributed thousands of cuttings to farmers who had never found a solution to the bank erosion of wandering streams and already we are getting back reports of success. They have gone even as far as Oklahoma where the boys from the Samuel-Noble Agricultural Foundation took them back by the hundreds. They can be well used in Oklahoma, for no state has suffered more damage from wandering streams and sudden floods.

We found also a benefit which we had not been looking for. The population of mink and muskrat has increased a dozen times since these water-loving animals find in the thickly tangled and matted branches along the stream banks exactly the kind of cover and runway they love best.

In the summer of 1949, one of the boys in a visiting Veterans' Class, observing the thick growth of willows and the fashion in which it had benefited our land and our game population, observed, "Sometimes it looks to me as if we'd got to go all the way back to the Indians and begin all over again to undo all the damage our grandpappies have done to this poor country."

Of course, what he was saying was that, so long as man works with Nature, she will reward him with whatever he desires, and that, when he works against her, he merely destroys his own land and cattle, and eventually himself. The Valley farmers discovered that fact after they had spent thousands of dollars to destroy what Nature had constructed and lost thousands of dollars worth of good land downstream into the Gulf of Mexico.

But the story of the little stream does not end there. As one by one the farms upstream put into practice programs of proper land use, the floods and the siltation began to diminish. The farms which did not put in such programs have since been abandoned or are on the way to total abandonment, for many of the farms in our Valley were on their last legs many years ago. Those fields which have been abandoned have grown over with weeds and underbrush and such a development is better for the land and for the whole of the Valley than the fashion in which those fields had been farmed in the past.

Once the abandoned farms are grown over with weeds and brush, the run-off water, the floods and droughts and the erosion are all very largely checked and the poor abused earth is given a chance to grow a healing blanket of cover.

Today the little stream is rarely discolored save in corn-planting time and, save for the floods of the Bad Year, 1947,[1] when there was so much rain (more than in the Great Flood of 1913) that the earth could drink up no more, the stream has never been out of its banks and rarely rises more than a few inches above its permanent level. Moreover, the flow has been stabilized. While there are no more sudden floods, the flow of clear cool spring water throughout the hot months has increased steadily with each successive year and is still increasing. As the springs throughout the watershed come back to life and flow once more, the game fish and the water vegetation are coming back and watercress actually grows again on the riffles.

The story of the stream follows very closely the story of the springs at Malabar. We are in a country of natural springs but when we came to Malabar there were not more than five really good springs that did not dry up during the hot summer months. Since we came to Malabar all five have increased their flow from 50 to 200 per cent. At least twelve springs, some of which I remember as good springs when I was a boy and which had dried up completely in summer during a generation, are flowing again, all the year round, and we have four good flowing springs where there were none at all within the memory of anyone alive.

The reason is simple enough and capable of explanation by any precocious kindergarten youngster. Instead of 85 per cent of the rainfall running off the acres of Malabar within twenty-four hours after a thundershower or cloudburst, carrying with it good soil, today we keep over 95 per cent of the rainfall where it falls, to sink into the ground and feed our springs. It was proper land use turned the trick. Each year we have more water. We have put in two new farm ponds on sites where only a few years ago there was no water at all and where now there are springs to feed the ponds. On the whole of about 1000 acres there is not a field today where we cannot turn cattle, give them a block of salt and forget them. We now have good water everywhere, a blessing which every livestock man will appreciate.

And there are other astonishing evidences of the rewards of proper land use and of working with Nature.

[1] Described in *Malabar Farm*.

[121]

On the Bailey place we constructed a dam to create a pond from the waters of the famous deep-level spring at the Bailey house. The site drains a watershed of about 110 acres, and in order to build the big dam we excavated with bulldozer and scoop to a depth of about sixteen feet. In the process we made an astonishing discovery. At a depth of about sixteen feet we found an old string of tiles, of the old-fashioned kind with holes poked in them to let the water through, a fact which indicated they must have been put there at least two generations earlier. Ten feet above we came upon another string of tiles, possibly put in a generation later. Above that lay another five feet of topsoil. All of it, to a depth of fifteen feet, had come off the Bailey hills on the watershed above where, when we acquired the place, there was no longer a single grain of topsoil.

Below the dam where a ditch had been dug years earlier to carry off the waters from the spring, the ditch was so filled with topsoil that its bottom lay more than two feet *above* the surrounding land. The overflowing water had turned an area of 4 or 5 acres of good land into swamp. When we dynamited the ditch to reclaim the surrounding swamp area, we uncovered an old farm bridge beneath six feet of eroded topsoil which had come down from the hills above. The dynamiting worked, and today we have an excellent truck garden on the site of the former swamp.

But the rest of the story is the best part. The Bailey pond is now five years old and save in the Bad Year, the water has never risen even to the top of the dam. There is no emergency flood outlet although the pond drains a watershed of 110 acres. A ten-inch pipe disposes of whatever flood water there is. On the very site where we excavated about sixteen feet of topsoil, the pond today remains perfectly clear after the heaviest cloudburst. In other words, if the land on the hills above continues to be properly managed, that pond (on a site once buried by ten to twelve feet of topsoil) will remain unchanged for a hundred thousand years without ever once being washed out or without so much as a grain of siltation. Again the change was brought about merely by land use.

The puzzle is that this factor did not become apparent generations earlier before floods, erosion and siltation had cost us as a nation billions of dollars in losses and ruined thousands upon thousands of once prosperous farmers. Both the evidence and the solution were there all the time. It merely needed a little observation and a little intelligence to correct the evils. Only a few naturally good farmers and the Amish and the Mennonites ever understood the calamity and how to deal with it.

In the beginning at Malabar, it was necessary to remake the whole pattern of the farm from the old-fashioned square fields on which erosion had been so devastating into curved fields and strips which ran around the slopes and hills. Instead of plowing up and down hill we plowed around the hills so that the furrows, instead of channeling the rainfall down the hills, aggravating old trunk gullies and creating countless new ones, ran nearly always on the level or something near to it. This was necessary for two reasons, because we were still growing corn which meant open row-crop cultivation and because on most of the farm the organic content was so close to zero that the cement-like quality of the soil merely turned the water off the surface instead of absorbing it like a sponge as any soil high in organic material will do. Moreover, the hillsides were laid out in alternate strips of sod and row crops so that, if the water carrying soil with it broke away from a strip growing corn, it would run into a strip covered with sod, which in turn would slow down and absorb the run-off water and allow the soil to be deposited in the strip of sod on our own land instead of in the Gulf of Mexico.

These were merely engineering measures, although fundamentally related to sound agriculture. They were, in a sense, emergency measures and did not necessarily mean that the land would leap at once into high production. The measures laid a foundation upon which we could rebuild a sound and profitable and permanent agriculture. As Dr. Kellogg of the Soils Service of the D.A.C. has pointed out, erosion is not a cause, it is a result of bad farming. After a few years, because we ceased growing corn altogether and because we had immensely increased the organic content of the soil so that it drank up the rainfall on the spot instead of turning it off, these strips, contours and curved fields were no longer necessary and were abandoned. In the beginning, however, we could not have done without them.

Emergency and mechanical as these measures were, they had an almost startling and immediate effect of checking the water run-off. They cut this run-off from about 85 per cent in a heavy rain to less than 20 per cent, and we were not the sole benefactors. Our neighbor, Charlie Schrack, much of whose land lay just below our own, reaped a lion's share of the rewards. Although Charlie was in the beginning dubious about the early goings-on at Malabar as crank and new-fangled, he became one of the earliest converts in the Valley. As he put it, "For eighty years my father and I have been fighting two big gullies that were made entirely by the run-off water from your land. Today no more run-off water comes down those gullies. I have plowed them in and am growing good crops right across them." Charlie's fields, too, changed their appearance in time from the old square fields into long strips which ran across and around his hills and, as a result, his crop yields, which had always been good because Charlie is a good farmer who understands the earth, increased more and more each year. He, too, kept the rainfall where it fell and his soil along with it.

In the beginning, near the Big House there were two mighty gullies, and during the summer months every thunderstorm brought a whole creek of water down each gully, across the barnyards and down to the creek below. So violent was the run-off that the only possible solution seemed to be a diversion ditch to catch that devastating flood halfway down the hill, subdue its speed and divert it, by means of the wide, shallow, level ditch *around* the hill, to run off over the heavy sod of a nearby old orchard. Herschell Hecker, the Soil Conservation Service engineer who helped us to lay out the whole farm on a plan of modern Soil Conservation farming, designed and built the ditch. During that first summer it served its purpose well, but from that summer onward there has never been a drop of flood water come off the land above even in the heaviest cloudburst, and the dry ditch has become known as "Hecker's Folly." It is there for all to see, its purpose nullified merely because the land above it was properly managed and the millions of gallons of good water which had once coursed down the steep slope are now kept on the land above to feed moisture to crops in late summer and replenish the reservoirs of the old dried-up springs. Hecker was right in setting up the original diversion ditch as an emergency measure; the subsequent development merely proved the always obvious truth that you do not stop floods at the mouth of the Mississippi or the Missouri but upstream in the forests, the cultivated fields and the grazing lands.

On the opposite side of the Big House, when we came to Malabar, there was a big ravine and a small flattish piece of land above it which for most of the year was a swamp. The ravine itself had been the product of long erosion and the flattish piece of swampy land above it had been, generations earlier, a still pond for the making of whisky. Long before we came to the farm the still pond had become entirely silted up with topsoil from the land above. We decided to dam the ravine and create a pond and to drain the swampy land, through which a small spring stream flowed which once fed the still pond. The plan was to turn the old silted-up still pond into a garden for shrubs and flowers.

During the first summer when we began these operations, cynical neighbors leaned on the fence and said, "You're wasting a lot of time and money. That little stream goes wild in summer every time there is a big rain and washes out everything. It will tear out any garden you make there. And in summer the stream goes dry and you won't have any water for the pond you're building."

The result turned out to be exactly the opposite. Only once or twice has the little stream ever threatened to make the least flood trouble. We grow iris and goatsbeard and other flowers down to the very edge of the running water and they have never even been damaged. As for the pond, it has never been short of water because, as we kept the flood water on the land above, it sank into the earth and increased greatly the flow of the springs which fed the little stream so that now it flows all through the summer, watering our cattle, our garden and constantly replenishing the big pond where we have sunfish, blue gills, bass and even rainbow trout in abundance.

At the side of the pond there still remains the foundation of the first cabin which had been built more than a century earlier over a good-sized spring so that in case of Indian attack there would always be water inside the house. For a generation or more the spring had gone dry every summer about the middle of June. Today it is flowing again as it did in the days of the first settler and provides a comfortable good flow of cold water right through the hot dry months.

In *Pleasant Valley* I told the story of the wholly dead spring that came back to life high up on the hills on the Ferguson place. Our only clue to the fact that there had once been a spring on the worn-out pasture in that particular spot was the presence in the field of old-fashioned garden flowers—red day lilies, Star of Bethlehem and the like. We knew from the flowers that, although no trace of it remained, there must once have been a cabin on the site because no settler ever built a cabin far from a good source of water in case of Indian attack. But the surrounding field was bone dry save for a little seepage during the early spring months.

In April of the fourth year, Pete Gunder, the herdsman at that time, came in one morning while I was having breakfast and said, "Come up to the Ferguson place. I want to show you something." He refused to tell me what it was he wanted to show me, but after we had climbed to the top of the Ferguson hill and gone part way

[125]

down the other side, we came to the spot where the old-fashioned flowers grew. There bubbling out of the earth was a clear cold stream of water where before there had only been a dry depression. The spring has flowed steadily ever since and feeds the farm pond constructed on the site and waters our cattle and the wild life throughout the hot dry season.

In the beginning the spring's only fault had been that it seemed to wander about a good deal, flowing one season in this spot, in another season perhaps thirty feet away. We only discovered the reason when we built the farm pond and unearthed the original spring. Not only had it dried up but the original source had been buried under two or three feet of topsoil from the hills above. On digging out the topsoil we found the original spring, walled in with stone by the hands of someone who had been dead for a century or more. With its source unblocked from silt, the spring no longer wanders but flows again in the same spot as when the first settler built a cabin there. The answer was very simple. By farming the land properly above the spring, we checked all erosion and run-off water and the accumulated rainfall gradually refilled the underground reservoir until it flowed again and increased its flow a little more each year.

The rediscovery of that spring was in a way an archeological operation like the unearthing of the ruins of Nineveh.

One of the best of recent developments in the New Agriculture has been the widespread increase in numbers of tanks and farm ponds. These are benefits to sportsmen, to the farmer, to wild life and in general to the conservation of both soil and water, for no farmer in his right mind, hankering for a farm pond that will water his stock and provide swimming and fishing for the whole family, would build a farm pond which will silt up in a few years because he has not properly managed the land on the watershed above it. To have a decent, clear, good pond or tank you have to keep the rainfall where it belongs and prevent the soil from washing away.

At Malabar it is a simple matter on the whole to build and maintain a farm pond for we have in our hilly country any number of small ravines and depressions which need only to be dammed and the springs, which increase in number and flow each year, provide the water supply. At present we have five ponds and others are planned. Indeed, Bob Huge says that if I continue building ponds, the farm will be entirely under water.

To construct some of these ponds requires only a couple of days' work with the Ferguson equipment of hydraulic-lift blade and scoop.

Once they are established and filled with water, a half hour's fishing in one of the older ponds provides a milk can full of bass, blue gills and sunfish with which to stock the new pond. Vegetation, both shallow- and deepish-water varieties, is transferred to the new pond to establish itself, and in a year or two we have a new fish pond with no further trouble than to fish it perpetually in order to keep down the number of fish.

The area about the pond to the size of an acre or two is fenced in to keep the cattle from trampling the dam and muddying the water and this area is allowed to go wild and is planted with all sorts of shrubbery—standing honeysuckle, hazelnuts, multiflora rose, viburnums and other food plants and shrubs—which provide both food and cover for rabbits, quail pheasant, squirrel, raccoon and other game.

In *Malabar Farm* I have devoted a long and detailed chapter to the variety of delights which these ponds bring to a farm. The ponds and the surrounding cover provide a constant increase in fish and game for the sportsman, fresh fish for the farm table, wild ducks and geese, swimming for the kids and the grown-ups in the hot dry months, and unlimited water for the protection of farm buildings in case of fire, water for the livestock and irrigation water for the vegetable gardens to guarantee them against any drought. And for those who desire simply to watch and observe Nature and wild life and the marvelous fashion in which Nature herself rapidly transforms a new, raw, freshly dammed pond into an old established one, the farm pond is a source of endless pleasure and fascination.

Not the least factor in the conservation of water at Malabar has been the treatment of the woodlots. There are about 160 acres of wooded land, most of which was pastured by sheep and cattle a couple of generations before we took over the land. The practice was poor management and the results in some parts of the woodland disastrous, for the animals found in the woodland very little to eat and what they did find had little more nutritive value than so much old wheat straw. Forced to eat anything at hand, they had largely destroyed all the protective covering of forest floor vegetation—the ferns, the wild flowers, and, worst of all, the forest seedlings. Some parts of the woodland were little more than a handful of big trees of doubtful value with low-quality weeds and sickly grass growing underneath.

Our predecessors at Malabar were getting from that woodland neither any decent forage nor any succession of good timber. Much of the land covered by trees is rocky and rough and of little value if cleared even as pasture land, so the trees were left undisturbed but the woodlands were fenced to keep cattle out except where shade and water and what my Kentucky friend, Forrest Borders, refers to as "tramping ground"[2] were left.

As to good forestry practices, we followed the advice of the Forestry people but not all of the way. The "ripe" trees, of which there were a surprising number left for some reason I have never been able to discover, were cut and used in remodeling the farm buildings. "Wolf" trees, or the trees which have no value but shut out sunlight and air and use up moisture and fertility, have gradually been removed. We have, however, permitted the old or dying trees to remain as dens for raccoon, squirrel and other wild life and we refused to clear out dogwood and wild grapes because they are excellent food sources for wild life and also add greatly to the beauty of the woods.

In the period of ten years since the cattle were first kept out of the woodland, a miraculous and beautiful change has taken place. The forest cover has returned with its carpet of ferns and wildflowers as well as thousands of seedlings of beech, maple, white ash, black walnut, hickory, tulip poplar and many kinds of oak. Some of these seedlings have already attained a height of twenty-five to thirty feet, according to the species, and all are straight-growing and smooth. Moreover, the wild-life population and the birds have increased their number many times. During that period we have cut "ripe" timber to the value of several thousand dollars to make way for the new seedlings, which one day will represent thousands of dollars worth of beautiful wood in a world badly in need of it. The whole crop may not be harvested during my lifetime but in the meanwhile we have back the singular, almost tropical beauty of the Ohio hardwood forests, an increase in game even to deer and grouse, and no water any longer runs off the forest floor. Even the old gullies have healed over as the run-off water from the fields above has been totally checked.

Below the largest piece of woodland lies a swampy 30 acres known at Malabar as "The Jungle," overgrown with underbrush and some valuable timber. It is ideal game cover and filled with springs and in May the earth is carpeted with trillium, violets, anemones, hepaticas,

---

[2] "Tramping ground" or rough ground with weeds, water and shade is of very great value to livestock and especially to dairy cattle and during hot dry weather can increase the milk flow and the health of the animals as much as an equivalent number of acres of good pasture. There is a good deal of space devoted to the subject in *Malabar Farm*.

Dutchman's breeches and other wild flowers. Here the springs flow more and more each year to join the flow of a creek which no longer floods and changes its course each year tearing out the old deep fishing holes. Very happily and very rapidly we are getting back to the Indians.

The only inconvenience arising from the control of land and water has been the appearance here and there near the bases of our bigger hill areas, sometimes as large as an acre or two in extent, of what might be described as seepage spots or "wet weather" springs. This inconvenience is small in comparison with the restoration of the springs over all the farm, and we have in most cases managed to correct the overabundant moisture merely by continuously plowing in heavy sods on these spots and greatly increasing the organic content. As we do so, the top level of the soil becomes each year more and more porous, the excess water draining down to the harder surface of the subsoil nine to ten inches below. As most of these seepage spots occur on our hill land which is kept largely in heavy pasture, they have never been a very great inconvenience, but by process of continually augmenting the organic material we have corrected the drainage in most of these spots, and in the hot dry weather they provide us with our best pasture. We might have tiled them out but it would have been an expensive process, with considerable upkeep. In our opinion at Malabar we would never put in a tile where natural drainage may be achieved in some other fashion.

We had two spots on the farm where it appeared unlikely that we could ever establish or maintain proper drainage without the use of tiles. They were both natural bowls surrounded by lips in the one case about twenty feet high. The larger bowl covered about 12 acres of our best bottomland. There was no outlet and the bowl drainage also suffered from the fact that the run-off water from the adjoining county road ran into it. The 12 acres lying in the bottom of the bowl never quite became a swamp except during the winter and early spring, but it was impossible to farm since it was always too wet to plow until about the first of July and it was too wet to grow proper alfalfa or grasses. If we plowed it when dry enough in August and seeded to wheat, the wheat was drowned out during the winter. The only possible solution appeared to be tiling, but the estimates were discouraging. It would cost about $2500 to cut through the lip of the bowl in order to get enough fall to carry off the water. And in any case we preferred to keep the water on our own land and we did not like the idea of draining our natural fertility and our bought commercial fertilizer down the tiles in solution with water. Beyond that, there was the tiresome and sometimes expensive upkeep of tiling.

The solution came out of my memories of farming methods in Belgium, France, Denmark and Holland, so one Sunday we hired in a bulldozer and scoop that worked during the week on rebuilding an adjoining county road, and went to work. We dug in the dead center of the bowl at the lowest spot a pit twenty by eighty feet and ten feet deep. The earth which we excavated (which was all topsoil from the adjoining hills) we merely leveled off on the surrounding area with the blade of the bulldozer. Then we planted fast-growing Baby-lonica willows along the edges both to evaporate the water and to bind the banks against crumbling frost action in the winter.

The results were exactly what we hoped. The excess water, which formerly soaked the better part of the 12 acres until well into the growing season, all drained into the pit, making a pond of it for the larger part of the year. The following summer we planted the whole of the bowl to wheat and in the following March seeded the area to alfalfa, brome grass and ladino clover. The wheat flourished up to the very edge of the pond and brought us a big yield and the field is now in a heavy growth of meadow. The alfalfa, which will not tolerate poor drainage, is flourishing. The whole job cost us $60 and an afternoon's work with bulldozer and scoop, instead of $2500 or more. We are keeping our water where it belongs and there is no expense of upkeep whatever, no clogging or breaking of tiles and for a part of the year at least we have another farm pond.

In the summer of 1949 we set up an experiment in irrigating permanent, rotated pasture and meadowland with the co-operation of the Aluminum Corporation of America, the Skinner Irrigation Company and the Gorman-Rupp Pump Company. We had long irrigated the vegetable gardens with excellent results, as to both quantity and quality of production, especially during the hot dry months. Irrigated from our shallower farm ponds, the temperature of the water was close to that of the air and the pond water carried a considerable amount of fertility not only in mineral but in organic form.

The irrigation of pastures and meadows in the Middle West, with its well-balanced and distributed rainfall, may seem to many farmers perhaps an exaggerated procedure, but the experiment was eminently successful and profitable possibly because the equipment used put on such large quantities of water in so short a space of time. Running two large overhead sprinklers, we were able to apply from an inch to an inch and a half of water every two hours over an area of 4 to 5 acres at the rate of near to 1000 gallons of water per minute, thus making it possible for one man to irrigate upward of 20 acres of pasture or meadow a day. The water came out of Switzer's Creek fed by the constantly reviving springs of the now well-managed farms upstream. The corner has been turned in so far as the volume of creek water is concerned. Ten years ago its volume was diminishing each summer. For the past three years it has been steadily increasing.

As to the results, we gained even in a summer of well-distributed rainfall at least 30 per cent in production on permanent bluegrass pasture and on the ladino pasture during the hot dry months as much as 60 per cent upward. In the meadows the second and third cuttings of hay were increased by at least 30 per cent since the brome grass and the ladino, both inclined to go dormant during the dry hot weather, produced prodigiously under irrigation. In a fairly long season the irrigation would make it possible for us to produce four instead of our usual three cuttings of mixed legumes and grasses. The gain was not only in quantity but in the milk-producing greenness and tenderness of the grass and legumes as well. We could not have done it without the water from Switzer's Creek which under a declining agriculture on the watershed above us *could* eventually have dried up altogether during the summer as more and more small streams have done in recent years throughout the United States because the water upstream was allowed to run off the land instead of sinking into it.

Not the least factor in the conservation of water, the checking of erosion and the rapid incorporation of quantities of organic material in the soil arrived when we abandoned wholly the old-fashioned technique of "clean" plowing and fitting in which our grandpappies took so much pride. The dangers of "clean" plowing and fitting of fields low or devoid of organic material was evident from the very beginning to anyone who took the trouble to observe what happened. As my friend, Dr. C. L. Lundell of the Texas Foundation, has put it so well, "pretty" farming has cost the nation hundreds of millions of dollars.

In the early days at Malabar even a field sowed to small grains eroded badly on the least slope or rise in a field for the very simple and obvious reason that on soil low in organic material "clean" plowing and fitting left the surface bare and hard with very little capacity to absorb water. In consequence the water simply ran off, creating gullies and carrying off more and more soil, both top and subsoil.

It was our roughest, abandoned land which gave us the clue to the virtues of "rough" plowing and fitting the kind of soil which we took over. In such fields, overgrown with broom sedge, poverty grass, goldenrod, wild aster and other coarse-growing weeds, it was impossible to "clean" plow or fit any field. There was only one way to subdue and bring into order those masses of coarse weeds and that was by ripping up the fields as best we could and then disking them thoroughly until the trash was chopped into the soil. On fields, prepared in this fashion, we discovered at once, there was zero erosion and zero water loss, for the soil, mixed with the rubbish, was so open and loose that all rainfall was absorbed at once.

We also discovered by observation another fact, vastly important to ourselves or to any farm in need of rapidly increasing organic material (or, indeed, to any farm at all). We discovered that in these rough-fitted fields, the rubbish plowed under—even the coarse woody stems of the goldenrod—disintegrated quickly into humus and became a part of the soil. On the "clean" plowed and fitted fields where the soil had been turned completely over, buried and compressed into a tight, thin, putrefying layer, it was possible three years later to plow up old cornstalks, straw and even traces of barnyard manure which had scarcely disintegrated at all. This occurred even in fields where comparatively heavy applications of nitrogen were made.

The reasons for this delayed disintegration of organic material into soil were obvious enough. By the overfitting of the soil above the buried trash, all air, including of course both oxygen and nitrogen, was excluded and on the "clean" fitted surface much of the water ran off and never seeped into the earth at all. With these important factors in the process of disintegration prevented from performing their natural functions in the universal process of birth, growth, death, decay and rebirth, the organic materials, from barnyard manure to old cornstalks, were merely buried, suffocated and in the end mummified.

Moreover, that tight, thin, impacted layer prevented the movement of water and moisture in both directions, upward and downward. It made a kind of layer of cardboard nine inches down which in wet weather prevented the water from seeping away and thus impaired drainage and left water standing on the surface or in the oversaturated top nine inches. Conversely, in hot dry weather the same layer of cardboard prevented moisture from rising by capillary attraction from the lower moist levels of the soil to feed the crops and provide the moisture necessary to making the fertilizer, both native

[132]

and purchased, available to the crops. More than this, such a system of "clean" plowing and fitting also accentuated the natural tendency of the mouldboard plow to create over many years the almost impenetrable hard pan which has increasingly reduced production on thousands of American farms and put many of them in certain areas out of production altogether.

Our first step was to stop "clean" plowing and fitting altogether, even upon stubble fields and good clean sods where rubbish was no obstacle. We simply ripped up the fields with the mouldboard plow and then disked them to a point where the drills or planters would cover the seed. As little rather than as much fitting of the soil was done as possible.

Gradually we have come to abandon the mouldboard plow almost entirely, using in its place the Graham-Hoehme plow or the Seaman or Ferguson tillers. The first is a kind of gigantic harrow with curved teeth which rips up the earth to a depth of fifteen to sixteen inches. The Ferguson "tiller" operates upon the same principle but is a lighter implement and needs less power to pull it. The Seaman tiller is an implement operating on the principle of the rototiller, but much larger, which literally "chews up" soil, sod, rubbish manure and other organic materials *into* the soil to a maximum depth of nine to ten inches. None of these implements tend to create a hard pan and the Graham-Hoehme plow and the Ferguson "tiller" do exactly the opposite. If used at a sufficient depth and with sufficient power, they shatter the old hard pan and restore drainage and consequently high production on soils which for years have suffered from a declining production owing to bad drainage, or have become actually waterlogged because the hard pan barred the water from seeping naturally into the lower levels of the earth. But both implements serve to mix all organic material *into* the soil rather than merely turning it over, burying it and compressing it into a tight layer.

In the case of these implements, the cost in man-hours, gasoline and depreciation of machinery is enormously cut because, by the time the field has been "plowed," half or more of the fitting has already been done, and in the case of the Seaman tiller we have put in many fields of small grains immediately after the tiller has gone over the field once. The Ferguson offset disk operates in a similar fashion and is one of the greatest of fitting tools.[3]

[3] Dr. Frank L. Duley, Professor of Agronomy, and J. C. Russell, Research Professor of Agronomy at the University of Nebraska, have together worked out a whole system of excellent farm implements from fitting tools through drills and cultivators for handling stubble or trash mulch farming in the Great Plains

The Seaman tiller is a remarkable advance in the field of agricultural machinery but it is at its best in the hands of the good and intelligent farmer who understands what good soil is and how it should be managed. In the hands of the "short-cut" farmer or the ignorant one, it can become actually a dangerous implement, for if used on soils low or deficient in organic material or without considerable quantities of sod, manure or just plain rubbish to incorporate in the soil, it may, by its rapid and rather violent action, disintegrate soil structure to a point where the soil becomes cement.

It is clear, I think, that no intelligent farmer would ignore organic material and that any intelligent farmer would understand the bad and wrong use of such an implement. The tool has been immensely valuable to us in the creation of the trash mulch seedings which have worked such miracles on the barren, weed-covered Bailey hills. Where once with field cultivators and disks it required at least six or seven fittings to incorporate the trash into the soil to a shallow depth and level it off for seeding, we are now able with the Seaman tiller to prepare the fields in one or at most two operations—an immense saving in man-hours, gasoline and in capital investment in many implements. The Graham-Hoehme plow and the Ferguson "tiller" and offset disk are, however, absolutely foolproof. Any farmer can use them only to the benefit of his crop yields and soils.

On some of the land at Malabar which was a rich but heavy clay, we went beyond mere rough fitting and employed a chisel to rip up the fields and break up the hard pans which on the clay ground had been created by a hundred years or more of mouldboard plowing done when often enough the ground was too wet to be plowed. One of the serious results of underground hard pans is that, as they come slowly into existence, the drainage deteriorates slowly. Each year the field dries out for plowing a little later and each year the farmer is inclined to plow it a little wetter in order to get in his crop. This process naturally only aggravates the conditions and hastens the process until at last the hard pan becomes impenetrable and the field will not drain at all. There are in the Middle West alone millions of acres of such land in existence today.

We had small areas of such pre-glacial clay in many of our fields. Actually these were small swamps in which marsh grass was the

---

area where the prevention of wind and water erosion and the conservation of moisture are absolutely necessary factors to a continuing agriculture in that region. Both have contributed enormously and almost alone to the development of trash mulch technique in the growing of corn.

principal crop, even on the high flat fields of the Ferguson place. These spots would grow nothing at all but coarse swamp grasses and it was impossible to get a seeding of good grasses or legumes because in winter the spots actually stood under water. We cured the condition entirely by going into the field during the hot dry months, when the clay hard pan was baked and hard, and breaking up the hard pan with a single deep chisel, sinking it to a depth as great as twenty-four inches. When this depth could not be achieved in going over the spots once, they were gone over a second time at right angles until the proper "cracking up" process was achieved. It was possible actually to *hear* the hard pan cracking and breaking beneath the power of the tractor and chisel.

The results were what we had expected. During the following winter *no* water stood on those clay spots. It drained deep down through the cracked hard pan into the earth. What we did not expect was a whole volunteer crop, appearing in the spring, of alsike, timothy, native white clover and even some ladino and alfalfa which had not hitherto germinated. We shall, of course, never again put a mouldboard plow on those clay soils, and now that it is possible with restored drainage to grow deep and coarse-rooted alfalfa and sweet clover on such soils, the question of a hard pan will not again arise. It was a simple process cheaply done and it has restored many good acres of our land to permanent high production.

At Wichita Falls Malabar in Texas we had during the summer of 1949 a perfect example of the immense value of chiseling and breaking up hard pans as a means of conserving and utilizing water. The soil at Wichita Falls Malabar had been badly used for years, overirrigated and plowed with the mouldboard or turning plow until the whole of the farm was underlaid at a depth of a few inches by a cement-like hard pan about two inches thick. Before putting in any crops, every field was chiseled and cross-chiseled thoroughly. Rainfall in the North Texas country is erratic and frequently comes in torrential downpours within the space of a very short time. Exactly such a thing happened during the summer of 1949 a little while after we had seeded about 100 acres to sorghum. On the chiseled field, a heavy two-inch rainfall disappeared deep down into the earth while on the adjoining unchiseled fields the water stood on the surface of the dead flat fields until at least 50 per cent of it was evaporated by the hot sun and the perpetual winds from the Great Plains.

This hard-pan situation is particularly true of much of the adobe-

[135]

like soils in the irrigated areas of the Southwest. In the Wichita River Valley and in many other similar areas of the Southwest, so much water has been poured onto the land that thousands of acres of once rich or potentially rich soils have been converted into marshland, a condition which has actually made necessary expensive draining and ditching operations in order to restore it to production. In other areas the wasteful and excessive use of water high in alkali content has put the land out of production altogether. Moreover, the excessive use of irrigation water simply aggravates the problem of hard pans until eventually little or no water penetrates below a depth of nine or ten inches but remains standing in pools on the surface to drown out crops until it is wastefully evaporated by wind and sun.

The abuse of irrigation water is a doubly serious problem in many parts of the Southwest where the water supply is limited and where agriculture and to some extent population are dependent upon reservoirs of irrigation water for their very existence. Parts of Arizona and Southern California have already reached that point where a decision must be made whether irrigation agriculture will be continued or increases in population and industry must be controlled or cut off altogether. In certain areas there is simply no more water either on the surface or underground and there seems to be no solution save the discovery of some reasonably economic method by which sea water can be reduced to fresh water or by the economically prohibitive project of bringing water into the area all the way from Washington and Oregon where there is a surplus of water for irrigation.

Much of the abuse and the waste in the irrigated areas arises from the ignorance of the individual farmer and his lack of understanding concerning soil and water. Great numbers of men have gone into irrigation farming, the most difficult and tricky form of agriculture, without ever having learned simply what good farming is or what constitutes good and productive soils. The result is that instead of farming by conserving moisture to the utmost degree they attempt to make water do the farming for them, in the belief that by merely turning on the water and letting it run they will get the results they desire and expect. In the dead flat land of the Wichita River Valley I have seen gullies twenty feet deep and half as wide cut out of the flat fields into the adjoining river bed which were actually created by flood irrigation water turned on by the farmer and allowed to run through the night.

[136]

Farther to the west, in the Texas Panhandle where irrigation water is largely supplied by artesian wells, the supply is giving out at an alarming rate of speed. When it is gone, that minerally rich Panhandle soil will simply become grazing or semi-desert land again since there is small means or possibility at present of impounding rainfall during the rainy months and using it for irrigation during the hot dry ones.

In the same area and, indeed, throughout the whole of the vast wheat-growing area of the Great Plains, the disastrous dust storms which plagued the nation as far east as the Atlantic coast during the Thirties were caused almost entirely by the hand of man and by a wretched wheat-growing, single-crop agriculture. Those wheat areas are very largely operated under a system of single-crop agriculture which requires the same or a greater need for specialized methods than irrigation agriculture.

In the old days in the country where wheat was grown year after year on the same land, the straw and weeds were simply burnt off to get them out of the way and prevent them from clogging the mouldboard turning-plows and the primitive harrows of a greedy, careless frontier agriculture based upon "clean" plowed and "clean" fitted fields (again a devastating example of "pretty" farming). Very rapidly this process of constantly depleting the organic material produced a soil which in dry weather was like powder and in wet weather like cement. It could not absorb the rainfall which merely ran off, creating gullies, carrying off good soil, and causing disastrous floods and siltation downstream. In dry weather, the soil, having nothing to bind it together, simply blew away in clouds so vast and heavy that street and automobile lights had to be used at midday. I have seen hard-surfaced roads in the Panhandle country standing five to six feet above the level of the surrounding fields from which that depth of soil had been blown or washed away in a period of a generation or less.

Today it would be difficult to find a mouldboard or turning-plow throughout the whole of the vast wheat-growing area from the Arctic Circle to the Gulf of Mexico and "clean" or "pretty" plowing and fitting is a thing of the past. So also are the dust storms and the violent water erosion of the past. Yields per acre have been rising steadily and a crop failure has been unknown since 1939 regardless of extreme variations in the amount and distribution of rainfall. The possibility of future dust storms on any scale even approaching the storms of the Thirties has been almost certainly eliminated so long as the present tillage methods are used and expanded.

This immense revolution has been almost entirely brought about in a comparatively short period of time simply by a technological change in methods of agriculture—a shift from the old-fashioned, careless, ignorant and devil-take-the-hindmost farming to the principles of the New Agriculture. Tree shelter belts and other devices have had some effects but wholly negligible ones in relation to the fundamental one of the disappearance of the mouldboard plow and "clean" plowing and fitting.

The mouldboard plow has been replaced almost entirely throughout the vast Great Plains area of the United States and Canada by one-way disks which chop the accumulated straw, weeds and roots two or three inches into the surface, or by the Graham-Hoehme plow which does a much better job by mixing the organic material *deeply* into the soil and by constantly breaking up old hard pans without any risk of creating new ones. It is, I think, only a matter of time until the Graham-Hoehme principle of fitting dry-land, wheat-country soils will dominate entirely, to the exclusion even of one-way disks.

Today the combines leave the wheat straw behind them on the fields. In most areas, the big Graham-Hoehme plows follow directly behind the combines, ripping up the earth to a depth of sixteen inches and working straw, weeds and trash *into* the soil itself. The field is left until the spilled volunteer wheat has germinated and the field is then gone over a second time with the Graham-Hoehme, either diagonally or at right angles. By this time the soil is open and loose to a depth of from ten to sixteen inches, and when the typical violent rains fall, every drop of water sinks deep into the earth instead of running off a dusty, "clean"-tilled surface. Moreover, the rubbish left on the top acts as an insulator against evaporation by the hot sun and by the hot winds which, in the Great Plains area, are even more destructive of moisture.

According to the Soil Conservation Service this rubbish, left on the surface, will reduce the velocity of a sixty-mile-an-hour gale to a velocity of eight miles an hour on the surface of the ground. Under such conditions (primarily the conservation and preservation of moisture) the question of insufficient moisture has largely been eliminated. In other words, an annual rainfall of eleven to twelve inches under the New Agriculture is the equivalent today for crop purposes of a seventeen- to eighteen-inch rainfall under the old "clean" plowing, mouldboard-plow agriculture. At least six to seven inches of rainfall does not run off the dusty surface and is not evaporated while standing in pools on the surface after a heavy cloud-burst. All of it goes into the ground. The element of erosion, either by wind or by water, is virtually eliminated as well.

To be sure, here and there, one discovers a bad farmer who still thinks that "what was good enough for Grandpappy is good enough for me," but very rapidly he is becoming eliminated, either by the ruthless law of economics and declining yields, or because his neighbors, unwilling to be suffocated by the dust or half-drowned by the gully-washing floods from his fields, have behind them the pressure of neighborhood public opinion and in Texas actually of the law itself. There now exists a statute in Texas under which the neighbor of a bad farmer whose fields are blowing or washing may give notice, and if the warning is not taken, the neighbor may go into the bad farmer's fields and "roughen them" up with implements of the Graham-Hoehme type to stop the blowing or washing.

This, of course, is the first faint foretaste of the kind of legislation which may be passed in every state or perhaps even in Washington, unless our agriculture improves more and more rapidly and our farmers choose to farm properly to correct such abuses as well as to augment their own incomes. In several states proposals have already been made by which a farmer may bring legal proceedings to compel a neighbor who turns silt and run-off water across his land to take the proper measures for the conservation of both soil and water. Some of the proposals make it possible for the complainant to sue and collect damages from the offender.

Of course the question of whether such legislation and even more severe legislation comes into existence rests with the individual farmer. At present many of them are asking for it and, unless conditions change, they are certain to get it. In England today, unless a man farms his land properly, the Government puts another man in his place who does a more intelligent job, or the farm is sold over his head and he is dispossessed. Such procedures are by no means impossible or perhaps so far away as they might seem in this country. Hugh H. Bennett, chief of the Soil Conservation Service, in an address at Princeton University when he was given a degree, stated the belief that the time might not be far off when a farmer, in order to be permitted to farm, would have to have a license to practice exactly like any doctor, lawyer, engineer or other professional man. All this is a long way from the abandonment of "clean" plowing and fitting at Malabar but essentially it is all merely a discussion of a fundamental principle related directly to good, living, productive soil and the elimination of water and soil losses. In the beginning some of the neighbors in the Valley chided us about not knowing how or being able properly to plow and fit a field. None of us at Malabar are particularly thin-skinned and the evidence of our eyes and of our farm account books was all in our favor and we continued in our own way, building soil fertility and yields, checking erosion and keeping the rainfall where it belonged. Gradually, over a period of years, we have seen a good many of our younger neighbors begin to follow our example, and some of the older ones have come in asking for the recipe by which we were able to turn abandoned, useless land like the Bailey hills in a short time into highly productive pastures and meadows. The fields at Malabar have always told their own story which to the farmer himself is more eloquent than anything that can be said or written.

However, to satisfy ourselves in every detail regarding the merits of "rough" fitting, we tried for three years strips of wheat put in side by side in the same soil with the same seed and the same amount of fertilizer. The only difference in the two strips was that one was roughly plowed and fitted, the organic material mixed into the soil and fitted as little as possible, and the other was fitted in the traditional way of our countryside in which the field was clean-plowed and then fitted again and again and again in the traditional way with harrows and culti-packers until the field resembled more a dusty county road than a field in which to grow crops. During a period of three years we averaged from 10 to 12 bushels per acre more wheat off the rough-fitted than off the "clean"-fitted strips. We knew already what the results were likely to be but the tests merely made the results indisputable.

The reasons again were clearly evident if one used his powers of observation. When I sought an answer to the reason why it was customary in our part of the country to fit any field for wheat by turning it into the consistency of a county road, I succeeded in getting only two answers, neither of them very valid. The first was, "Because that's the way we've always done it" which had no validity whatever, and the second was that the fields were treated that way in order "to maintain the moisture."

Observation showed us that this was exactly what the "clean" fitting did not do. When the first rain came after the wheat was sowed, the big raindrops, falling on the dusty "clean"-fitted surface of the field, merely splattered out and sealed up the dusty surface, turning most of the rain off the strip or into the lower portions of a field or down the nearest stream. When at last the soil dried out, it left a hard, caked surface. On the rough-fitted strip the action was exactly the opposite. Because the soil was loose and porous and filled with rough organic material, every drop of rain was absorbed and penetrated deeply into the earth, and when the surface dried off, the soil was still loose and open and sponge-like, with no resistance whatever to the tender young shoots when they tried to find their way to the light.

When after two or three weeks we dug up certain of the wheat plants from the two strips, the root system of the wheat on the rough-fitted field was at least twice as deep and as diffuse as that of the wheat on the "clean" strip. The reasons were childishly obvious. In the loose-fitted strip the soil was open and porous and there was no resistance whatever to the penetration of the tender, fast-growing rootlets. In the "clean"-fitted strips the soil was so hard and packed that the rootlets had to fight their way. Moreover, the loose-fitted strip remained well aerated and encouraged that diffuse and hardy growth of roots described elsewhere as surrounding any air pocket in the soil. The difference was also evident in the deeper greenness and vigor of the wheat on the rough-fitted strip both because it had the bigger more vigorous root system and because the moisture content beneath the surface was much higher and therefore made the fertilizer more quickly and more completely available.[4]

[4] Both at Malabar Farm in Ohio and at Malabar in Texas we have had extraordinary examples of the effect of aeration on deeper soils and of the vigorous mineral fertility of deep subsoils when brought to the surface. At Texas Malabar many of the fields were leveled for irrigation purposes and banks built around them from the excess soil. In many cases all the old topsoil was scraped off altogether and the actual subsoil was used in the construction of the banks and borders. Volunteer wheat left from a preceding crop germinated both on the old topsoil and in the scraped-up, aerated subsoil. The difference in the wheat plants in the two soils could only be described as fantastic. The wheat growing in the aerated subsoil was deep green, vigorous and produced at least three times as many stems in stooling out as the yellowish sickly wheat growing on the old topsoil. In Ohio when the gas company excavated to a depth of three to four feet in order to lay a pipe line, the subsoil churned up to that depth by the operation produced a similarly prodigious growth of wheat in comparison with the old topsoil alongside. Most surprising of all was the evidence in the growth and color of the wheat plants of great quantities of nitrogen in the subsoils which were virtually devoid of any organic material. The only explanation of the phenomenon was given by an agronomist friend who held that the nitrogen in the tight subsoils excluded from light and air existed in the form of nitrites unavailable to plants in that form and that exposure of these nitrites to air and light transformed them into highly available nitrate form.

The same difference was observed in the spring when the wheat plants began to stool out. The stooling was much heavier on the rough-fitted strips, the stalks four to five inches taller and the heads decidedly bigger and better filled.

During one of the three winters in which we had perpetual freezes and thaws, there was no loss from "heaving" on the rough-fitted strip while our losses on the "clean" hard-fitted strip were upward of 25 per cent. Again the reasons were evident. The hard, "clean"-fitted strip rose and fell under the freezing and thawing in a single sheet like an asphalt road, lifting the wheat out of the ground and leaving it suspended there with half the roots exposed. On the rough-fitted strip the earth, mixed with organic material and rubbish, merely crumbled and left the plants still firmly rooted. But the effect of the differences in fitting did not end there. They extended into the second year when the seeding of alfalfa came on. The losses from "heaving" of alfalfa on the "clean," hard-fitted strip were considerable. On the rough-fitted strip there were no losses whatever. On the rough-fitted strip the germination was at least 25 per cent greater with no "washing about" into the low spots whatever.

All of these are the reasons why at Malabar the mouldboard plows, covered with grease, remain longer and longer periods in the machinery shed, until, as Jim Cook, boss of the field operations, observed, they were beginning to become "antiques." No one had ordered them out of the fields and no one has kept them in the machinery shed on order or because he was trying to prove something. They were left there in favor of the Seaman tiller, the Graham-Hoehme plow and the Ferguson offset disk and "tiller" because these implements do a much better job, give us bigger yields and, under some conditions, they also cut down at times by as much as 80 per cent the expenditure in man-hours and gasoline given over to fitting a field. And certainly they are the answer to preventing water loss in cultivated fields and to a complete check of all soil erosion either by wind or by water. Every year they make our topsoil deeper, improve our drainage, break up old hard pans and prevent the creation of new ones. The process also follows our principle that the more machinery is kept off our fields the better.

In *Plowman's Folly* Edward H. Faulkner first brought the trash mulch method to wide popular attention. He made, I think, two errors which later he largely admitted in his book *A Second Look:* (1) He insisted that trash mulch farming was the *only* method of good agriculture, which is certainly not true. (2) He did not go far enough.

He insisted that shallow trash mulching was enough and our own experiments both in Ohio and Texas and the experience of the big Great Plains wheat growers have refuted this. At Malabar we would say that in our experience the trash mulch system is a great step forward in agriculture but that shallow cultivation and mulching is not enough. The deeper you go within limits, the better the results with regard to moisture, soil and water conservation, increased organic material yields, mineral availability and weed infestation.

None of what is set down above is any longer either theory or experiment. Malabar Farm has been in operation long enough for us to have obtained tangible, practical, inarguable results. In all respects, from beauty to dollars and cents, the change has been enormous. Indeed, the very landscape itself has changed so that parts of the farm and even some of the smaller views no longer bear any resemblance whatever to the original landscape we took over. Each year the land, saddling the watershed at the lower end of Pleasant Valley, becomes greener and more comely, with better crops and more and bigger springs, less water loss and erosion. Some of the fields are today giving us bigger yields than the same land ever did as virgin soil and today the acreage supports more people than ever before in its history at a much higher standard of living. I think we have partly accomplished what the G.I. suggested—"that we go back to the Indians and begin all over again to undo what our grandpappies did before us."

And Malabar is by no means the only farm where such a regeneration has occurred. It has happened and, for the good of the nation, is happening daily to thousands of other farms through the Middle West, the East and the South. There is nothing very expensive or very difficult about such a transformation. It is all part of the pattern of the New Agriculture made largely here in the United States and based upon the eternal pattern of a good agriculture which has not varied since Adam was driven out of the Garden of Eden and put to work . . . which was probably the best thing that happened to him or could happen to any of us. In Ohio at least the New Agriculture has made so much progress that it is no longer the subject of mockery as "new-fangled." The shoe is on the other foot and will be increasingly so as the younger generation, trained in 4–H Clubs and the Future Farmers of America and the Vocational Agricultural classes, comes along and takes over our good earth. "What was good enough for Grandpappy" isn't good enough for them. In fact it was never good enough for anybody.

[143]

CHAPTER VIII

The Muskingum Story (a note related to *Water and the Farm*)

The grass is rich and matted. It holds the rain and the mist and they seep into the ground feeding the streams. . . . It is well-tended, and not too many cattle feed upon it; not too many fires burn it, laying bare the soil. Stand unshod upon it, for the ground is holy, being as it came from the Creator. Keep it, guard it, care for it, for it keeps men, guards men, cares for men. Destroy it and man is destroyed. . . .

—ALAN PATON, Cry, the Beloved Country

VIII. *The Muskingum Story* (a note related to *Water and the Farm*)

IN THE year 1913 the state of Ohio underwent the greatest disaster in all its history, an event which has never been forgotten by any of those old enough to remember it. One April morning it began to rain, not merely a good heavy rain, but a downpour described by the natives as "raining pitchforks and hayladders." It rained all that day and the next and the next and the next, steadily and violently. This was no sudden, violent but brief cloudburst such as the West and the Southwest sometimes experience. It resembled more closely the forty days and forty nights of rain described in the Old Testament which caused Noah to build his Ark despite the jeering of his neighbors.

For days the heavy rains continued without a break until even on the watershed of Ohio, in the highest part of the state, cities like my home town of Mansfield were flooded. Trains on the transcontinental lines ceased running, powerhouses were flooded and cities were left without light or power. Flooded sewage plants polluted the water supplies and residents of cities were forced to drive into the country to springs for safe supplies of water. Bridges were washed out and farmhouses swept away by such currents as the oldest inhabitant could not remember, while towns and sections of towns were submerged up to second-story windows. When the rains finally ceased and the waters had gone down a little, it was discovered that more than five hundred people had lost their lives and more than $300,-000,000 of property damage had been done. This became known, and will remain known so long as there is a state of Ohio, as the Great Flood of Thirteen.

Now Ohio is a state unused to disasters. It knew periodically the flooded lowlands along the Ohio River and its tributaries. People had grown accustomed to them, but the citizens of Ohio had no experience with wild tornadoes or forest fires or earthquakes or the periodically disastrous floods familiar in some regions. They were resentful and indignant that Ohio should be visited by such a catastrophe. It was an indignation of almost comic proportions which James Thurber has described in "The Day the Dam Broke," one of the finest pieces of humorous writing in American literature. It is a short story about the Great Flood of Thirteen and of the bewilderment and indignation it aroused in the bosoms of the citizens of Columbus, Ohio.

Fortunately the indignation did not die away. The citizens, especially in the areas which suffered most, decided that something must be done about it. After a long waiting period in which there was much dissension, something was done. The citizens of the area which had suffered most in the valley and watershed of the Muskingum River, the biggest of Ohio's in-state rivers, finally drew up a plan known as the Muskingum Watershed Conservancy District. It became a reality in 1933 and since then has operated consistently and brilliantly to the benefit of those in the watershed itself.

The immediate area was organized as a public corporation with the power to plan and to construct and administer flood control and conservation projects, issue bonds and levy assessments and taxes. It also possessed the power to enter into contracts with the Federal Government or the state of Ohio for co-operation in any project undertaken. Making decisions for the corporation were a board of common pleas judges and a board of elected directors. Behind it all there stood the figure of one man by the name of Bryce Browning who, at one time during the Great Flood, had been working with the Zanesville Chamber of Commerce and witnessed the death and damage in the Muskingum Valley. The plan was largely his conception and it became his obsession. For nearly twenty years he fought doggedly to keep alive the indignation of Ohioans and get them to "do something about it." Today he is secretary and director of the Muskingum Watershed Flood Control District. He lives for it and probably will die for it. His conception has created for the great Muskingum watershed and for the people of Ohio a kind of Paradise of Recreation and it has stopped dead in its tracks the threat of future death and disaster from flood in an area covering 8,038 square miles of rich and beautiful Ohio land. It has created handsome forests and helped vastly to check all the destructive erosion which was eating up rapidly some of the finest agricultural land in the world. It has created a necklace of twelve beautiful lakes as lovely as any to be found in the English Lake Country. There is boating, bathing, fishing, camping and hunting close at hand for the Ohio millions who live in great industrial cities like Cleveland,

Youngstown, Akron, Toledo, Columbus, Cincinnati, Dayton and many other smaller industrial communities. City dwellers anywhere in Ohio need drive at most only a couple of hours to find beautiful lakes and some of the finest fishing in the world.

Ohio is, surprisingly, second in fur production among the states and territories, and much of its fur is contributed by the marshes and wild country of the Muskingum Conservancy District where wild game, muskrats and even mink abound. Each year more and more wild ducks, many of them of the deep-water variety, follow the flyway created by the twelve lakes which join up, below the Ohio River, with the ladder made by the great lakes of the Tennessee Valley Authority area. There are no more threats of property damage and loss of life. But most striking of all, the whole District is the only area and project of its kind which for maintenance and continuance costs the taxpayers nothing at all. It supports itself and pays acre per acre the same taxes as the other land in the big watershed.

This all came about because the plan was conceived broadly, not merely as a power project or to prevent floods alone. Conservation of soil, water and forests, increased property values, particularly in the cities, better agriculture, recreation, development of wild life and the planting of forests and rental of land for revenue were all taken into consideration. In short, a whole watershed through the center of one of the richest of our states was developed as an entity with consideration being given for every aspect of its potentialities and it was not done by government and government withdrew or canceled no tax revenues.

There is no longer any doubt about the workability and the great benefits of the pattern. The District has been in operation long enough to prove itself. In the spring of 1947 rains almost as heavy as those of the Great Flood of 1913 descended on the Valley[1] (indeed, in the upper reaches the downpour equaled the violence of the 1913 flood rains) and not one cent of property damage occurred nor was one life lost. The fourteen sluices of the great dams were closed and the flood water impounded and held there, to be let out slowly over the succeeding weeks. There was no need for expensive levees which in the past often broke and created more damage than if they had never existed. There still remained in each of the great reservoirs enough reserve space for flood waters twice the size of the 1913 flood, a situation which is virtually inconceivable.

[1] A description of this disastrously wet season is contained in the author's *Malabar Farm* in a chapter entitled "The Bad Year."

The interest of the Federal Government in such a project was undeniable since the prevention of floods and siltation in the big Muskingum watershed was not only of immediate value to the area itself but also of great value to the areas bordering both the great Ohio River and the mighty Mississippi all the way to the Gulf of Mexico. The Federal Government provided an original fund of $22,500,000 for the construction of the upriver dams and those on the Muskingum. The state of Ohio voted a fund of $2,000,000 for the acquisition of the necessary land. The operation of the dams and gates themselves was left in control of the Army engineers who still operate them.

The Conservancy District itself works in close co-operation with both federal and state agencies having to do with all phases of flood control and conservation of natural resources. These include the Division of Forestry, the Soil Conservation Service, the State Commission of Natural Resources, the state and federal Agricultural Extension Services, the state Highways Department, the state Division of Wildlife, the National Wildlife Service, the Geological Survey and many private organizations having to do with all phases of conservation. On the borders of the District at Coshocton, Ohio, exists the largest hydrological and training school station in the world dealing with the problems of water in all its phases. All of these agencies contributed much to the establishment of the original plan of operation and continue to make valuable contributions. Perhaps the most notable element in the operation of the area is the freedom from friction, feuds and jealousies among all these agencies, perhaps because of the wisdom and the tact of Mr. Browning, the secretary and director, and the board of over-all directors.

It is notable that the pattern and the idea is spreading within the borders of Ohio. On the Miami River in Southwest Ohio a series of dry flood-control dams already exists, although the basic plan offers nothing approaching the comprehensive developments of forestry and recreation provided by the Muskingum District. Both the Muskingum and the Miami Rivers are tributaries of the Ohio, but on the Maumee and the Cuyahoga Rivers which flow into Lake Erie, plans are already under way for the establishment of watershed conservancy districts on the pattern of the Muskingum District. Similar plans are also under way for the Hocking River, a smaller tributary of the Ohio.

This development within the borders of the state of Ohio offers a solution, and a very nearly perfect one, for the troubles which afflict the watersheds of most states in the modern United States. Other states are beginning to follow the plans. Texas has set up in the Trinity River watershed a control plan extending far upstream to the headwaters which is as good for that area as a dozen gold mines or any number of new oil wells. The plan of the Muskingum District can be repeated almost everywhere, by the citizens of the area themselves. The only exceptions are in sparsely populated wild areas such as those existing in the West and in the mountainous parts of the South.

In the past taxpayers, property owners and insurance companies have spent billions of dollars in repairing damages and in constructing and maintaining and repairing vast dams and levees at the *mouths* or halfway down our great rivers, when all the time it must have been easily evident even to a kindergarten child with a pile of sand and a watering can that one does not stop floods at the bottoms of rivers but high up on their tributaries and in the forests and cow pastures. Despite this obvious fact, virtually the whole of our flood-prevention work has been concentrated until now upon huge and vulnerable dams and levees far downstream and pretentious but often foolish reservoir dams and huge diversion canals. The inefficacy of such action has been proven again and again in our history. On the lower Mississippi dikes and levees now carry the muddy waters as much as forty-five feet above the level of the surrounding territory and still that territory is constantly subject to the breaking and washing out of the expensive barriers and to the greater damage which occurs when a dam or levee breaks and releases water suddenly with a wild, ferocious violence. Even now, Army engineers are planning and building dams for the Missouri Valley and other areas and expensive levees and barriers on the lower reaches of that big, turbulent and muddy river. As certainly as the sun rises tomorrow, those great dams will fill up with silt and become useless and the great levees will break and release augmented destruction. It may be fun for the engineers to build such dams and levees and it may make a lot of money for the contractors on construction and eventually on the highly profitable repair bills, but such tactics are immensely expensive to the already long-suffering taxpayer and for the property owners of the areas affected. At times it would seem that the motto of the Army engineers should be "not efficiency but size and expense."

During the spring floods of 1947, the Muskingum Conservancy District dams proved completely the value and the efficacy of these upstream, tributary dams. Not only did the dams prevent the loss of a single life or a cent of property damage within the watershed area, but they also impounded an immense volume of water which did not add its burden to the already flood-devastated areas of the Ohio and the Lower Mississippi. Damage on the Ohio and the Mississippi that year ran into many millions of dollars. The loss would have been doubled and more if the flood waters of the Muskingum watershed had been added to the downstream torrents.

The operation of the Muskingum Flood Control District is really as simple as A-B-C. If, for example, similar flood control districts existed on the upper tributaries of the whole Mississippi watershed, the flood waters of that whole vast area could be impounded and released slowly when the flood dangers had passed or be used for irrigation or for creating electric power. The cost would be infinitely less than the construction of vast levees downstream which periodically break with great damage and loss of life and of the huge dams which after a few years fill with silt and become useless. Moreover, the construction of these smaller dams upstream, very often in rugged territory, decreases enormously the artificial flooding of valuable agricultural land and the destruction of whole prosperous communities by the vast and ineffectual dams and lakes constructed downstream.

It is possible to imagine a Mississippi watershed, much greater than most of the countries in the world, which would in time be free of virtually all flood danger, with an immense saving to the nation in taxes and damages and immense conservation of water for other purposes.

Since the construction of the T.V.A. dams and those in the Muskingum district, the flood waters of the Ohio River, of which the Tennessee River and the Muskingum are both tributaries, have been cut in half. Likewise the burden of flood waters pouring into the Lower Mississippi. Suppose the other tributaries of the Ohio—the Allegheny, the Monongahela and a dozen smaller streams were treated in the same fashion. The periodic flooding of the Ohio would be ended, and the flow of the Ohio, which lowers dangerously during the summer months so that dams and locks have had to be constructed to permit navigation, would be greatly augmented as the flood waters impounded by the dams were slowly released.

In time there would be created, in the areas surrounding the upriver dams, areas for forests, fishing and hunting, boating, swimming and other recreations such as those which exist already in the Muskingum area. These, as in the case of the Muskingum Conservancy District, can be self-liquidating and eventually actually profitable while paying their way without any burden upon the taxpayer or withdrawing any land from the tax duplicate as federal- or state-owned property. Indeed, the property values in cities and towns within the Conservancy and especially along the waterfronts have increased steadily. In many of the towns on the lower Muskingum there grew up during the past seventy-five years two towns—the old town built along the waterfront and a new town built on higher land above the level of the recurrent floods. In the course of the years the property values in the old lower town sank almost to nothing. Today with the flood peril completely abolished, the property values of the lower town have rapidly returned, both as residential and industrial sites.

But the Muskingum Conservancy District has not only prevented damage, it has created and creates increasingly new wealth in many ways. To the small towns in and about the lakes and forests there has come a big revenue from tourists and from sportsmen who visit the area to fish and hunt. The reforestation program, which continues steadily, is creating millions of dollars of new wealth and considerable employment. The large area of agricultural land owned, managed and rented out by the Conservancy to neighboring farmers is increasing steadily in production and yields. This is so because the agency leases the land on a yearly basis and exacts from the lessor a program of the best agricultural practices. If the renter violates any of these practices, the lease is simply not renewed and the land goes to a farmer who does a better job.

Of course, the forestry and agricultural program reaches much further than mere facts and figures. By the practice of a proper agriculture and the reforestation of bare eroded slopes, the amount of flood water and erosion coming from *within* the Conservancy watershed area has been immensely reduced and all erosion and the consequent siltation of the lakes has been very nearly corrected. Within a very few more years, the run-off water and the erosion factor in the area will have been reduced virtually to zero.

The influence of such improvements extends, of course, far beyond the borders of the Conservancy District, for their value is witnessed by the farmers of the whole big watershed and there is no doubt that agricultural and forestry methods throughout the greater part of Ohio are being greatly influenced for the better by the widening ripples of example and information coming from the Conservancy District itself. Slowly the whole big watershed, in the past badly farmed, eroded and with thousands of acres of land actually abandoned, is being reclaimed and transformed, and as the water and topsoil is kept where it belongs, even under the heaviest downpours, the flood damage, the siltation and the necessity for using the dams for impounding water each year grows less. Except for the flood year of 1947, few of the dams have actually impounded water in any considerable quantity during the past five or six years. This follows the pattern which proves again beyond any argument whatever that floods and erosion are not stopped at the bottoms of rivers but upstream in the fields, forests and cow pastures.

Some of the wildest and most beautiful country in the United States is to be found within the Muskingum Conservancy District. Some of this was once farmed-out, abandoned land which has been reforested, and some of it was simply wild, rough country, so wild that in certain areas it was completely inaccessible except to an occasional fisherman, hunter or woodsman. Much of this land still carries its virgin forest covering. It has now been opened up, discreetly, with good winding roads leading to canyons and ravines and inlets from the lakes where one can enjoy solitude as in the depths of wildest Canada or the Rockies, although he may only be an hour or two or even a few minutes from some of the greatest industrial cities in the world.

Nothing is more satisfactory than in the evening to come upon an industrial worker, who has knocked off work at four, and brought the wife and kids out into a primeval wilderness for picnic supper and some fishing. Each evening the road which passes through our own Malabar Farm is bright until long after dark with the lights of the cars bearing fishermen and their families home to the cities after a long summer evening of fishing, boating and swimming. Sometimes the car tows a small motorboat on a trailer, to be kept in the town garage until the next fishing expedition. Always the car contains a happy, healthy, sleepy family returning from an outing in some of the most beautiful country in the world.

The sources of revenue in the district are many and varied. They come from long leases on cabin sites deep in the woods, from concessionaires who rent boats and fishing tackle or conduct sightseers through the area, from the rents coming in from leased agricultural land, from small admission charges for upkeep of special camping areas, from the increasing revenue of the rapidly growing new forests. And perhaps the biggest chunk of income is derived from the leasing of fishing rights to the state of Ohio. In return the Conservancy has its own staff of fish and game experts, which practices the finest of fish and game management and provides some of the best fishing to be found anywhere in the world. Best of all, dance halls, honky tonks and saloons are prohibited throughout the whole of the area.

The sociological values are, of course, immense in terms of health and recreation and outdoor life and in developing the kind of citizenship which the United States must have if it is to survive. It is always possible for any family to escape the crowded quarters of our industrial cities for a wild and beautiful area within easy driving distance. Indeed, many workers and their families have set themselves up in cabins, small houses and even farms close to the Conservancy area and live there the year round, the wage-earners driving back and forth to work.

All the following factors should be borne in mind: (1) The Muskingum Conservancy District has established a pattern, too little known, which could be applied to the upper watersheds of most rivers in the United States, a pattern which could save hundreds of millions of dollars a year in damage, many lives, and actually create new wealth in dollars and cents and immense wealth in terms of health, recreation and good citizenship. (2) It creates no permanent tax burden to be added to that which is already breaking the back of our American economy. (3) It does not, as in the case of so much of our federally and state-owned land, withdraw from tax revenues great areas of land and then tax the citizens to pay for withdrawing it from taxation. The Muskingum Conservancy District actually pays taxes on every acre of land within its borders, at the same rate as that imposed upon any citizen of Ohio. (4) It belongs to the citizens of the area and not to an artificially contrived bureaucracy in far-off Washington, constantly at war with the citizens of the area.

The successful working model is there for anyone to see, right down the middle of one of the most heavily populated and industrialized states and one of the richest areas of its size in the world. It is worth going far to see, for it contains some of the most beautiful scenery in the nation and some of the best fishing. It is the nation's best answer to those who believe that the state should own everything and that all government should come from a patriarchal bureaucracy located in Washington. I know of no better example of real working American Democracy and its superiority over the National Socialist pattern.

At Malabar we live on the edge of the whole beautiful area, which in summer is a kind of Paradise. We rent one of its many farms. From one of our own high hills we overlook the beautiful Pleasant Hill Lake and several thousand acres of the Conservancy District.

[155]

We can testify that it is perhaps the greatest and most rewarding single project ever undertaken by the state.

Today, in the vast Missouri River Valley, there is a veritable warfare in progress over the fashion in which the flood and siltation problem shall be handled. The muddy Missouri offers the greatest problem of any watershed in the United States. For years it has carried billions of tons of silt along with its wildly destructive flood waters down the lower Missouri into the Mississippi and into the Gulf of Mexico. Since the Louisiana Territory was acquired, it has been clearly evident that dams and levees far downstream were futile measures in the control of the flooding rivers of that area. In a century or more, billions of dollars of taxpayer's money have been spent in building dams which only silted up and downstream levees which only broke and had to be repaired over and over again. Thousands of lives have been lost and billions of dollars worth of property destroyed, all because the simple and perfectly visible evidence that floods are stopped upstream and not downstream has been persistently ignored by those who have had charge of the flood prevention of our watersheds.

In the Missouri watershed, the same simple, stupid pattern is being repeated to a large extent in the plans of the Army engineers. Many of the projects advocate the construction of vast dams downstream which will eventually either break out or fill with silt, and the same projects call for the flooding of hundreds of thousands of acres of excellent agricultural land and the annihilation of whole prosperous communities. In the history of the world there have been few plans representing a more gigantic waste and futility. The object of flood and siltation control is not to build the biggest earthen dam in the world, not to construct the highest dam in the world, but to stop floods and siltation and conserve water, none of which is accomplished by the expensive and futile dams and levees built at the mouths of rivers.

It is reasonable to say that if all our watersheds were controlled as the Muskingum watershed is controlled, the flood and siltation bill for damages paid by the nation would be reduced by at least 75 per cent, and there would be little need for the hundreds of millions spent annually in repairs and dredging and upkeep.

The pattern is there, a pattern which any child can understand. Perhaps it is too simple and obvious. More likely it is simply not grandiose enough or futile or expensive enough to merit the interest of the military and the bureaucrats.

CHAPTER IX

A Streamlined Farming Program

Inside the barn no longer looks like a fortress. . . . Darkness surrounds us, with slits of pink light from the sunset filtering through the windows and the cracks in the walls. The light falls upon baled hay and alfalfa and the forms of the binder and the tractor. Let the sun be higher in the sky, and it would cast bands of deep gold across the heaped corn in the crib, and enrich the dull red of the piled cobs; it would dazzle in a confusion of small suns upon the metal of the machinery and slip its light beneath the doorway of the barn, to burnish the combs of stray chickens that scratch for grain on the haymow floor. Unwittingly we tread softly, for there is hay beneath our feet. But something makes us lower our voices in speaking, as though we stood in a holy place.

—CLARE LEIGHTON *in* Give Us This Day

IX. *A Streamlined Farming Program*

JUST outside Ithaca, New York, within sight of the towers of Cornell University lives a gentleman called Ed Babcock whose name at least is known to every good farmer in the United States. He lives at a farm called "Sunny Gables" and farmers from all over the world are likely to drop in at any time to see him. Undoubtedly he knows more about practical farming than any man in the United States. He is no longer a young man and his health is not the best, but he is always ready to talk farming and attack the special problems of farmers anywhere. He is tired from having spent almost the whole of his life in the service of farmers—in bringing about improvements in mechanization, helping to organize the co-operatives which have largely put an end to the ruthless exploitation of farmers in most areas. His latest exploit, and a very valuable one, is to work out sound plans by which a man can farm single-handed 120 acres or more of land, plowing, fitting, harvesting and putting up silage and hay for livestock. Each article, each book he writes is a gold mine of information and ideas. He is the originator of the idea, indeed of the plan which could well solve our surplus problems on a sound basis. Roughly speaking, the plan advocates a great increase in livestock and poultry under efficient management to consume the surpluses of grain from our single-crop areas and turn that abundance into reasonably priced, high-protein, "ice-box" foods within the reach of small-income consumers.

I have listened to Ed Babcock for years, and we write letters to each other quite frequently, principally I think because we are both interested in making more money for the farmer and bringing down the price of food to the consumer by making agriculture more productive and reducing costs of production by high yields per acre, by efficiency and by the use of brain instead of brute strength.

I know Ed Babcock lies awake at nights thinking how this can be accomplished and, above all, how the information can be given to the farmer and how more intelligent and modern programs for individual farms can be put into operation.

A chance remark of his crystallized for me a situation which lies at the root of many of our agricultural troubles. It was: "Too many of our farmers work twice as hard as they should to make half as much money as they should." The thought might still have been carried one degree further by saying that too many of our farmers work themselves to death to make no money at all.

Ed Babcock has not only been a great teacher and the farmer's greatest friend throughout his life but he has also, like the crew at Malabar, farmed in a practical way, finding many of the answers the hard way and always in terms of the average farmer and not of the "estate" farmer or the big show-ring livestock people. The hard way is sometimes hard, but when one learns that way, the knowledge is likely to stick and it can also lead on and on into new knowledge.

The farm at Malabar was set up originally on a four-year rotation general farm basis with the idea of self-sufficiency in all things largely dominant. The four-year rotation included wheat, oats, corn and hay and was accompanied by a livestock program including a few dairy cattle, a few beef cattle, hogs, sheep and poultry. It was a conventional rather than a radical or experimental program. It was a program advocated by all Middle Western state agricultural colleges and undeniably it had its virtues, particularly during the nineteenth century when there were few if any telephones or fast freight trains and no automobiles and no airplanes and distribution of farm products was on the whole limited and indeed primitive. It was a program evolved, partly by agricultural experts who sought to arrest the soil-robbing cash-crop agriculture which until quite lately has been virtually the rule in American agriculture, and partly out of the tradition and necessities which surrounded a primitive frontier agricultural life in which farms were isolated and it was necessary for the farmer to raise virtually all his food but sugar, salt and spices, and even to provide clothing for his family off the land on which he had settled.

It took us only about two years to discover that to a large extent it was a program that was out of date.

I confess that the self-sufficiency angle had a deep, traditional and somewhat romantic appeal for me and, during the war, when poultry, meat, cream and butter became rationed, scarce or at times unattainable, the program paid off well, for throughout the war the families at Malabar lived high, wide and open in so far as food and good rich food was concerned. It paid off even in the maple sugar and honey which took the place of sugar. But those were conditions of wartime when to some extent all of us lived under frontier conditions and they were by no means normal.

The four-year rotation general farming plan stumbled along fairly well for a couple of years and then slowly we began to realize that we were not making much headway toward the goals which we had set: (1) the restoration of fertility and production to the worn-out land; (2) a farm program which at some future time would bring in good profits with a minimum of labor and production costs and a maximum of production and profits; (3) our rotation program, with one year or two of grass, was designed to *maintain* organic material rather than to *increase* it rapidly, which was needed on most of our land which resembled concrete more than living and productive soil.

We discovered many things about such a general farm four-year rotation program. We found that we were doing a great deal of futile running around and that it took an excessive number of man-hours and different conditions of fencing and housing to maintain any considerable numbers of livestock of such variety. The dairy barn was no good for hogs nor would the ordinary fences contain them. The chicken housing required a great deal of remodeling and the construction of range houses located during the summer months in the fields. The sheep had a tendency constantly to migrate through fences which were not too good, always toward the high ground which was usually our neighbors' territory. The vegetable gardens took a great deal of time and labor so that, to produce enough in a common vegetable garden for all the families on the farm, the costs in man-hours were so high that very often we could buy vegetables out of the farm income more cheaply than we could grow them, and thus save money. The two-hundred-tree apple orchard which we inherited willy-nilly was not an asset in a world in which apples could be bought cheaply and could only be produced economically and at a profit if there were five thousand trees instead of two hundred.

And we found that under such a program we had the choice of investing many thousands of dollars in machinery, completely out of line with the income, or of having most of our work, and particularly harvesting, custom done, a process which is always unsatisfactory and at times expensive enough to eat up the smaller margins of profit. Such a program not only demanded the usual plows, harrows and other fitting tools. It demanded a tractor seeder, a corn planter and row-crop cultivators, a drill, a combine, a hay bailer, a corn picker and a great variety of other smaller tools. In order to compete at all in modern agriculture upon any but a subsistence basis, all these things were necessary. We might have harvested in the old-fashioned way by hand and by team, but we should have been working from dawn to dark at an immense cost in man-hours and always with the prospect of getting crops in too late because of the weather or of losing a large part of the harvest for the same reason. Moreover, in the first few years, the yields per acre at Malabar were too low to merit any such machinery. As production per acre and consequently profits increased, the bigger yields moved nearer to justifying such an expense in machinery.

In other words, it is my own belief that the four- to five-year rotation general farming plan is becoming obsolete. It was necessary on the frontier and it was justified during the period when agricultural experts sought to break down the old greedy cash-crop pattern of farming, but it no longer fits into the pattern of our times in a country which is highly industrialized, in an agriculture which, potentially at least, is highly mechanized, where distribution is widespread and at least mechanically efficient, and where total self-sufficiency is an anachronism and perhaps even an economic error.

Of course in all of this I am speaking in terms of the practical dirt farmer and businessman who wants good profits, a good education for his children, leisure and time to travel, short hours and high profits; there are some people who like and prefer the general farm, self-sufficient pattern of farming, as if they were still living on a frontier, because it provides them with a sense of security in time of war or of economic collapse. For "the farm as a family way of life," I feel a great sympathy and secretly I have a romantic attachment to the diversified, rotation, self-sufficiency school of farming, and it is an excellent pattern for the city office or industrial worker owning a few acres of land beyond the city limits, but I believe it has very little to do with progressive and highly profitable agriculture and has little or no place in modern agriculture. The farm wife who can raise fresh peas in her own garden but buys them in cans is a damfool and a woman with no taste for good food, but that is another matter involving good living and good food, and at Malabar we raise a wide range of vegetables which would perhaps cost less in the open market but would be inferior in flavor and quality to our own vegetables fresh out of the garden. There are romantics who are willing to accept long hours, hard work and low profits and believe that the mental peace they get from such a life justifies these things, but I do not believe they should be included in the category of a modern or even a sensible agriculture. Of course the old cash-crop farmer who grows corn or soybeans or cotton year after year on the same land is not a

[162]

farmer at all. He is merely a greedy ignoramus who eventually destroys his own capital and runs himself out of business.

What we discovered, the hard way, was simply that we were trying to farm during the twentieth century in terms of the nineteenth when both labor and time were cheap and little planning was necessary, when machinery was inexpensive and limited, and when there still existed two doctrines which were essentially and profoundly false: (1) that anyone could be a farmer; (2) that in order to be a good farmer a man had to kill himself with work.

The first doctrine—that anybody can farm—has cost this nation billions of dollars and produced in its day the "hick" who became a symbol of all farmers and the butt of vaudeville jokes. It destroyed hundreds of millions of acres of land for which we are paying every day in higher and higher food costs and higher and higher taxes. As to the second, the good modern farmer is the man who uses his brains and the knowledge and machinery available instead of his muscles. The brains and the knowledge mean high profits, high living standards, pride and self-respect and dignity.[1]

All this is a part of the revolution in American agriculture which has been going on unnoticed until very recently for three-quarters of a century. As a nation we have reached that point in agriculture which Europe, a far older country, reached long before us, a point at which there is no place for a pioneer frontier agriculture nor for the poor, wasteful and greedy methods of such an agriculture, but where agriculture is a business, a science and somewhat of a specialty. This fact, in short, is what we had to find out the hard way.

In the beginning the program at Malabar had been determined somewhat by the character of the land, the low yields per acre and the

[1] Today labor is never cheap except in a relative sense and one thing is certain—that with expensive machinery and valuable livestock the one thing which modern agriculture cannot afford is cheap labor. The only exception perhaps lies with the labor employed to dig carrots or harvest lettuce on the vast vegetable farms in Texas, Florida and California which are not farms at all but really industrial vegetable factories. And even under such conditions cheap labor, such as that imported during the war from the West Indies, can be expensive. On some of the great ranches of the South and Southwest, representing investments in land and cattle of millions of dollars, I have seen the management and work left to ill-paid, third-rate men. The cheapness of such employees or the "economy" represented by their low wages is a disastrous illusion, for in lack of initiative and unwillingness to accept responsibility they cause losses which represent sometimes as much as a hundred times the low salary they are paid. These cases are a supreme example of a penny-wise, pound-foolish policy in agriculture and livestock management that is common today among absentee landlords who not only incur heavy expenses in operation through such a policy but are also destroying the land which is their capital.

problem of getting some income. We carried some two hundred head of sheep to utilize the poor hill pasture and because we had a lot of land on which none of the forage merited buying steers to run on it. Most of the pasture was too poor to feed steers profitably, and we bought shorthorn cows which we crossed with Angus bulls to breed us blue-roan calves which we put on what good pasture we had after they were weaned. We know now what every good farmer knows or must find out—that poor land and poor fields never make any money for anybody no matter what the program or the dodge attempted. The sheep did fairly well and the shorthorn cows gave enough milk on poor pasture to bring up good, hardy, cross-bred calves which utilized the moderately good pasture probably better than pure-bred calves would have done, but they didn't make us any money principally because the forage was so poor that it required too many acres to carry them, and because the nutritive quality of the forage was inferior and the animals were never in the same prime condition of health and vigor as those we run today.

This was our first discovery of one of the fundamental rules of agriculture—that the livestock farmer, with the possible exception of the poultryman, makes his money in the fields and not in the barns. The profit he makes upon his milk or his beef or his turnover of animals is 90 per cent determined by the amount of grass, of grain or corn he produces per acre. The remaining 10 per cent is made by efficiency of operations, sensible balanced feeding and economy of man-hours. His livestock are merely the factory which turns the raw material, in terms of grain and forage, into dollars. The more he raises per acre the less this raw material costs him and the less he raises the more it costs him, not only in terms of taxes and interest, but in man-hours, in seed, in gasoline, in wear and tear of machinery and in the health and breeding capacity of his animals. Poor land produces not only poor people but poor forage and poor livestock as well. It is undoubtedly true today that most of our farmers are farming 3 to 5 acres to produce what under better practices could be produced on 1 acre.

We know now that if we had borrowed the money, limed and fertilized the whole place, put it down into legumes and had then begun our livestock program, we should have lost a great deal less money, made a great deal more rapid progress and saved ourselves a great deal of both heart- and backbreak.

However, we did not do so for two reasons: (1) We were and always are primarily interested in the problems of our average fellow farmers and especially those of the farmers dealing with the problem of the restoration of run-down land and we wanted to work it out in the terms of the farmer who has to pay taxes and interest and finance his farm as he goes while he restores the land and raises the yields per acre. (2) There was really no information available at that time on how to restore such land as ours, which had a deep and minerally rich subsoil with the topsoil either totally eroded away or so depleted and devoid of organic material that what fertility remained in it and even the commercial fertilizers we purchased were largely unavailable to the crops.

It should be pointed out that the project at Malabar was established about eleven years ago and that much progress has been made generally since then in the knowledge and techniques of land restoration, but we had to find out the hard way. We had to discover that much of the advice given us was quite wrong; for example, that alfalfa is a pampered crop and would grow only on rich loam, that it should be seeded in August or early September and that clean plowing was a necessity, and countless other bits of superstition and misinformation current at that time and still current in many areas and schools.

We had also to learn that corn was not a more profitable crop than heavy legumes and grasses but generally speaking a less profitable one, particularly in our area. The concept of grass farming was scarcely mentioned eleven years ago and the development of grass farming as a highly profitable method of agriculture has grown simultaneously and in a parallel fashion with the development of Malabar. We had to learn the wonders that trash mulches can accomplish in the restoration of land utterly devoid of organic material and the tremendous efficacy of organic materials in restoring our good drainage even without the use of tiles. Virtually no mention had been made of these things and many others only eleven years ago. At that time it was even the fashion to run down the great virtues of ensiled forage and on many farms sturdy silos which had cost much money were standing empty and beginning to rot away. Trace elements and their effects upon the health of plants, animals and people were virtually unheard of.

Moreover, we had the problem which faces any man attempting to restore worn-out eroded land in rolling or hilly country: we had had first of all an engineering job to do—that of taking out the old square fields and their fences, or laying out the slopes in terraces, diversion ditches and strips and doing our plowing along the contours of our hills rather than up and down the slope.

[165]

In those first years even the poor pastures, overgrazed in the past, were, in spots, so devoid of vegetation that the water and soil ran off under any heavy rain and the fields were so depleted of organic material that we could not even grow a crop of small grain on a moderate slope without risking serious erosion. We grew corn on strips and on the contour. We know now that it was economically foolish for us to grow corn at all, even on carefully contoured strips or even on our comparatively level ground, but we did not know it then and there was no one to tell us.

Corn and Middle West were terms which were inseparable. Corn was sacred. Corn unfortunately still is King in the minds of many Middle Western farmers just as Cotton was once King on the dilapidated lands of the Deep South. In our area a farm without corn was inconceivable. It was supposed that virtually no animal could do without it and that it was impossible to fatten a steer or a hog upon anything but corn. Yet corn was destroying the good earth of millions of acres of once good Middle Western land and countless farmers were actually raising corn at a loss and would have discovered the fact if they had ever taken the trouble to estimate the costs in terms of man-hours, seed, fertilizer, gasoline, taxes and interest, and the price they received either for the corn itself or the cattle they fattened on it at any but the highest yields per acre. And they overlooked entirely the slow destruction of the soil, which is the farmer's working capital, in terms of erosion and the destruction of organic materials and soil structure which arises from the constant plowing, stirring and fitting of soils which go with row crops.

Eleven years ago if we had not raised corn at Malabar we should have been looked upon as insane in our area or the neighbors would have said that we were too damned lazy to plow. Today more and more neighbors are following our pattern and by so doing are increasing their profits, reducing work drudgery and costs and building up the capital which is their land.

As a nation we inherited corn from the Indians, and it became, under a poor agriculture, like cotton, one of the most destructive factors in the general decline of our productive soils. I should be the last to advocate the total abandonment of corn which has a sound place in certain livestock operations and has a growing and eventually an even more important place in chemurgy for the production of alcohols, plastics and countless other products, the development of which is still in infancy. But the once common belief that corn is the most profitable of crops and indispensable in any livestock program and the indispensable backbone of our Middle Western rural economy is the sheerest hogwash.

Unfortunately it is economically a very destructive hogwash. It was once believed, and practice appeared to back up the belief, that only corn and hog- or beef-feeder farmers could be rich farmers and that therefore flat deep corn land was the most profitable land and brought the highest prices. Fortunately this kind of reasoning is beginning to change partly because cash corn-crop farming has virtually destroyed by erosion and depletion some millions of acres of good rolling land and run its owners into bankruptcy, and partly because we are beginning to find out that a well-managed farm program can be just as profitable without any corn whatever and is a much more stable operation since the costs are infinitely lower in terms of labor and fertilizer, and the prices for dairy products and to some extent beef are far more stable and dependable than the prices of corn and hogs. Moreover, even in many areas of deep rich soil, yields have been steadily declining and good drainage disappearing through an agriculture based upon corn, cotton, soybeans or similar row crops.

Corn, like wheat and cotton and tobacco, is one of the principal sources of our national surplus problem and of the burden that falls upon the neck of the taxpayer who must pay out millions in good tax money for support prices allotted, generally speaking, almost entirely to single cash-crop agriculture.

The reckoning in regard to such a policy is certain to come one day (and a day not too far off) because there are in the United States only about eighteen million farmers, including all the people who own as little as 3 acres of land or produce as little as $350 in total crop income per year and there are about 133 millions of other citizens who are growing weary of paying high food prices and, on top of that, hundreds of millions to subsidize in one way or another the single-crop cash-crop farmer in order to keep consumer's prices high. The city population and the general public are beginning to find out about such things, and the strongest farm bloc, even by log-rolling in Congress, will be unable eventually to pass on to the city consumer the burden of a greedy, ill-managed and generally poor agriculture in the form both of high prices and taxes. When single-crop subsidies are removed we shall begin to have a good agriculture, self-supporting and independent, which will stand on its own feet, conserve the fertility of our most important natural resource and make a far greater contribution to our national welfare and economy than our agriculture is making today.

One of the most interesting comments on "corn" agriculture and civilization is supplied by C. L. Lundell, an authority on the Mayan civilization of Yucatan and at present director of the Texas Foundation at Renner, Texas, set up primarily to find the means of restoring the blackland area of Texas, despoiled by row crops in a couple of generations. Concerning the effect of a single-crop corn agriculture in the history of the Mayan people, he writes:

When the Maya civilization developed in Peten and the adjacent lowlands of the Yucatan Peninsula, during the centuries preceding and immediately following the birth of Christ, the region was densely forested. With the increase of population, as evidenced by the numerous ruins, repeated clearing and burning of ever-increasing areas of the forest took place.

In the beginning the almost limitless virgin forest permitted the continued clearing of new land, from which high yields of corn, comparable to those of our Midwest, were obtained, and the surplus fed an ever-increasing population and provided the basis for the growth of classes, and the development of the arts and sciences.

To the early Maya, utilizing stone axes and fire for land clearing, the *milpa* system was a boon, but the system, dependent so largely upon fire, contained in it the element of destruction which led eventually to abandonment of the great cities, such as Tikal and Calakmul, mass migrations, and the ultimate disintegration of the civilization.

Under the *milpa* system, dependent upon fire as the clearing agent, surface litter is destroyed as well as much of the humus, causing fundamental changes in the chemical and physical nature of these soils. With the continued loss of humus with each clearing for corn production, and the break down of the physical structure of the *rendzina* soil through the destruction of organic matter, the tempo of soil depletion increases and erosion is accelerated. On such a depleted soil, the regeneration of forest cover is slower after each clearing, leading in time to reduced yields, requiring larger and larger acreages to support each population center. Concurrently, erosion leaves less and less soil with each season.

Considering the physical environment of the Yucatan Peninsula, a civilization built on a shifting type of agriculture, and centered in a one-crop economy, was foredoomed by soil depletion and the resulting erosion. The story is clear for everyone to read in the fastnesses of Peten. The uplands slowly were denuded of soil which filled the vast lake basins and lowlands with silt. Today the great swamps, known as *bajos*, surrounded by low limestone hills almost barren of soil, are testimony, even after a thousand years of abandonment, to the disastrous effects of their poor agriculture. Within the silt layers filling these swamps is the irrefutable and tragic story of the rise and fall of the highest civilization developed on this planet by aboriginal man.

The abandonment by the Maya over a period of centuries of their ceremonial centers, one after another, the repeated migrations of the Itzas, with ultimately a shift of the population to peripheral areas, is testimony to their fierce struggle for subsistence. In fact they practiced land rotation, and even rotated their areas of occupation according to nature's dictates, and in the end resorted to elaborate terracing in the southern peripheral provinces, but the disintegration of the empire could not be stemmed in a land depleted and eroded by one of the most destructive systems of agriculture ever evolved by man.

All this may seem a long way from a "streamlined" farm program but it is essentially a part of the whole picture. In one sense the record of Malabar has followed almost exactly the advances made in soil conservation, in grass farming and in other phases of a new and improved agriculture during the past decade. At Malabar we have worked as a kind of concentrated laboratory where these things could be studied at close range, in the fields, in the crops, in the livestock, in the farm account books and in the profits and taxes. We discovered many things from experiment and observation which had not before been tried or discovered or, if they had been, remained unknown to the average farmer. Many of these things have since become accepted practices and are coming into more widespread use.

As I wrote earlier, we discovered very shortly that we were spending too much money under a general farm four- or five-year rotation plan on capital investment in buildings and machinery and that under a rotation which included only two years of grass and legumes, even with extra rye and sweet clover cover crops plowed in, we were not making enough progress toward restoring organic material to soil almost wholly depleted of humus. (I do not believe today that one or two years of grass is enough to *maintain* fertility and organic material even in the North and Central states where it is burned out much less rapidly than in the warmer regions.)

Gradually we began to produce less corn and more good grass, and as we did so, the rising profits began to show up on the farm books principally in the money we did *not* spend, the man-hours we did *not* need and the machinery we did *not* have to buy. For three years now we have not raised any corn at Malabar and it is unlikely that we shall ever again raise corn. For one year we have not raised any wheat. Our only present rotation is from heavy meadows of brome grass, alfalfa and ladino clover into spring barley or oats and back again into alfalfa, brome grass and ladino.

Slowly, we have moved into a streamlined farm program, and as we have done so, labor and fertilizer costs and man-hours have declined rapidly. So have feed costs and capital investment in machinery and fancy buildings. Our carrying capacity, winter and summer, of cattle has increased from the original 30 head which could barely get through with the feed raised in the first year or two, to a capacity of from 275 to 300 head, and we are still 40 per cent from maximum optimum production and carrying capacity because much of our recently acquired land is only beginning to approach the productive capacity of land which we have had five years or more.

Most important of all, we have been able to prove that an acre of our land sown to alfalfa, brome grass and ladino brings in the same gross in terms of cash as that of the 90- to 100-bushel-an-acre corn farmer, and less than 5 per cent of the farmers growing corn in the United States raise as much as 100 bushels of corn to the acre. Moreover, our net profit in cash terms is at least two to three times per acre that of the corn farmer since the heavy grass costs us approximately one-seventh in labor and gasoline and one-fifth in fertilizer as against the labor and fertilizer of the corn farmer.

In the capital sense the gain is immense, for each year we are improving our soil and its organic content rather than tearing it down. Each year our soil becomes more productive and more valuable. Even upon our steep hills we have a zero record in erosion and less than 5 per cent water loss in the heaviest cloudburst. Each time we plow in a deep, heavy sod, each time we spread a load of barnyard manure, our land goes up in value, until today we could sell, on a Federal Land Bank appraisal basis, very nearly any of our land for three to four times its original value.

We make a claim at Malabar that our profit on a dairy animal or on a can of milk is, conservatively speaking, at least 30 per cent and upward greater than that of the average general farm rotation, corn-silage dairy operation in the huge Middle West. We can substantiate the claim very easily, especially in view of the figures uncovered in the recent survey of the New York milk-shed made by Cornell University which showed that the average profit of the dairy farmer was in that area one-quarter of a cent a gallon. If we made no more than that we should give up the whole business and go into something else. I will try to set forth some of the reasons for our claim.

At Malabar we buy no nitrogen fertilizer because our legume sod program produces excess free nitrogen for any crop save possibly corn. Although we achieved yields of 60 bushels per acre and sometimes more on some of our land, we have been driven out of wheat production because the excess free nitrogen produced wheat four feet tall which lodged badly, was difficult to harvest and ruined our grass and legume seedings, which we look upon as more valuable than the wheat itself. Even with high phosphorus and potash fertilizers, we cannot produce wheat without serious lodging. We are able to grow varieties of stiff-stemmed oats and spring barley in good yields, without lodging, and the grasses in our meadow, mainly brome grass and bluegrass, go wild from the abundant nitrogen and after a few years crowd out the legumes. Since nitrogen is in one sense only protein in another form, our forages are all high in protein-feeding values.

We buy no protein supplements or concentrates at $100 to $120 a ton since we manufacture sufficient protein for dairy cattle and beef feeders out of sunlight, air and water through our rich and heavy legumes. Except for bone meal, fed as a precaution to the cows being milked, we have bought no minerals for the past two years since apparently the cattle are getting all the minerals they need out of the forage from our *deep-rooted* legumes and grasses and seldom approach the boxes where minerals are fed at free choice. We have fed, until this year, no grain to any animal on the farm except the cows in the milking-parlor and then only a maximum of 5 pounds a day to the heaviest producers.

All of this money which we do *not* spend simply shows up on the profit side of the ledger. We could eliminate none of these expensive items in a dairy program based upon a four- to five-year rotation and upon corn silage or upon a program of timothy, red clover or even that green gold, ladino, for all of them are shallow-rooted, live and feed upon the top few inches of our soil, in many cases farmed year after year for more than a century only eight or nine inches deep, and do not penetrate into the vast mineral reserves of our deep and minerally rich subsoil.

Our grass silage costs us one-seventh in labor and one-fifth in fertilizer what corn silage cost us in the past. The same acreage of grass as of 100-bushel-to-the-acre corn will fill a silo with grass silage, and after the grass and legumes are ensiled, we still have two cuttings of excellent hay close in quality to alfalfa leaf meal. We shall have in the silo 25 per cent at least more roughage feed because the grass packs so much more tightly and the quality of feed is much higher than corn silage. And by experience we know that we take from at least 5 per cent upward more milk off the grass silage than off the corn silage.

It is necessary, I think, to explain the low cost of grass silage as against corn silage. We run the heavy mixtures of brome grass, alfalfa and ladino on the same land for a minimum of six years, some of it as long as ten. Those fields are plowed again only when the brome grass, going wild on the free nitrogen created by the legumes, finally crowds out the legumes themselves. During that period of six years upward the *only* labor and the *only* fertilizer expenditures are the top dressing of the field once every three years with 250 pounds of 0-12-12 or 0-20-20 commercial fertilizer, or a top dressing of barnyard manure. We have some seedings producing heavily after six years on which the only labor has been riding the manure spreader for about fifty man-hours over the field twice. The expense of bought commercial fertilizer on those fields is zero. We cannot fertilize or manure the fields more often because the grass and legumes become too heavy, lodge, are difficult to harvest and spoil our second cutting. Contrast this expense in fitting, seed, labor, gasoline and fertilization with that of the same field plowed, fitted, seeded, heavily fertilized and cultivated every year in corn for grain or silage.

It should be pointed out, as another means of measurement, that if we had the same acreage in corn as we have in grass and legumes we should have to employ a minimum of twelve men and twelve tractors at a very conservative estimate. Until the beginning of 1950 we have run Malabar and a fifty-cow milking-parlor with four men and five tractors and three high school boys who come in during the summer to make grass silage and bale hay and straw. If we produced corn silage to fill our seven silos we should have to have at least twice as many men and tractors, plus the hard work of making corn silage. When we went out of corn altogether after gradually cutting the acreage year by year, we sold one tractor, one corn picker, three cultivators and half our plows and harrows and fitting equipment—a capital investment representing about $5000, in addition to all the other machinery required in a four- to five-year rotation general farm program.

Our gross return on the year has more than doubled as we moved into a specialized grass-legume-livestock program and the profits on any crop have risen on a similar scale and a good deal more than that on milk and cattle.

Timothy we regard as a weed at Malabar not only because of its low feed value but because of its shallow rooting habit, and blue or

June grass and red clover are not much in favor for the same reason. Our imported deep-rooted legumes and grasses not only produce many times the yield per acre in quantity but many times the value in minerals and proteins and general nutritional values. We do not seed any timothy and red clover but we do have some 60 odd acres of bluegrass on land which is rough or on the wet side or intercepted by ditches and creeks which make the use of most machinery difficult or impossible. So long as we must grow bluegrass on such land, we came to the conclusion in the very beginning that it would be profitable to get from that land the maximum optimum production both in quantity and quality. We decided to treat the bluegrass and white clover areas not, as many farmers treat them, simply as wasteland which is only good for pasture, but as crop land raising a pasture crop which could be highly valuable.

The record of treatment on blue grass pastures over a period of ten years runs as follows: 4 tons of ground limestone put on in two applications at five-year intervals, 200 pounds of superphosphate the first year, 200 pounds of 3-12-1 applied five years later, and two top dressings of barnyard manure. Perhaps most important of all is the clipping of all bluegrass pastures at least twice and occasionally three times a year. The moment the bluegrass begins to grow tough and go to seed it is clipped as short as a lawn. When later in the summer it starts to seed a second time, the process is repeated and in a long season a third clipping is made. Each time it is clipped the clippings are laid down as a mulch over the sod, protecting the moisture from evaporation and maintaining a cool temperature at the roots. But the effect is cumulative as well as immediate, and after ten years, our bluegrass and white clover is growing through a mulch of organic material an inch or more in depth. When one walks over it, the effect is of walking upon a sponge. We find that bluegrass-white clover pasture does not naturally go dormant during the hot months; when treated thus, it averages a carrying capacity of about one cow per acre from early April to November.

When we took over these pastures they were typical of the old-time bluegrass pastures of farmers who believed that because animals dropped urine and manure on permanent pastures the fertility of the soil took care of itself. By July these pastures were six feet deep with ironweed, thistles and other weeds with a little mangy, dried-up, bluegrass through which the sunbaked soil could be seen.

The farmers who believed or believe that bluegrass pastures take care of themselves overlooked two important factors:

[173]

(1) Animals return to the field in manure and urine less than one-fifth of what they consume and that on those pastures (called "virgin soil" because they had never been plowed) hundreds and even thousands of pounds of phosphorus, potash and calcium, together with other minor elements, had been carried off the fields in the form of milk, bone, hair, wool and meat during the one hundred years or more they had been pastured. In many respects, the soil of those pastures which had never known a plow was more depleted than some of the heavily farmed fields, since at least the latter had had from time to time some applications of commercial fertilizer.

(2) We were aware that both bluegrass and white clover were shallow-rooted plants and did not have the power of penetrating deep below the layer of exhausted topsoil to find the deep minerals as did the deep-rooted grasses and legumes. Therefore they had to be fed directly by the use of commercial fertilizers or barnyard manure.[2]

By treating the bluegrass-white clover pastures as a crop, the carrying capacity was increased over a period of years as much as nine times. When we took over the land those pastures could carry at a conservative estimate one head of cattle to 6 acres for about six months of the year. Today we carry approximately one head to the acre on bluegrass right through the season from early in April until the snow flies. In the spring of 1949 we carried from April twelfth to June fourteenth over 50 head of milking cows on approximately 28 acres of bluegrass and white clover and they did not manage to keep up with the growth. On June fourteenth we opened the rotation pastures adjoining the bluegrass bottom.

These are pasture lots of from 8 to 10 acres each kept in heavy brome grass, alfalfa, ladino mixtures and rotated. Usually we first take off them a first cutting of grass silage, and by the time the carrying

[2] Undoubtedly the taste of cattle for such deep-rooted weeds as yellow docks is related to the presence of minerals originating from the subsoils which would not be present in shallow-rooted plants rooting only eight to nine inches deep. In *Malabar Farm* I have related the story of the dairy heifers which pestered our operations on opening a gravel pit in an alfalfa field. They hung about getting in our way in order to pick up the crowns and tough roots of the dislodged alfalfa which they chewed as a child would chew taffy. Undoubtedly they liked the high concentration of minerals they found in the roots and crowns. A neighboring old-time farmer pointed out to me that many of the deep green spots on worn-out bluegrass pastures were not caused by the urine or manure dropped by the animals but by the decay of the roots and crowns of the tough, deep-rooted ordinary bull thistle. These plants penetrate very deeply, are biennial and die the second year, leaving behind considerable organic material and very large amounts of minerals in highly available organic form which they have brought to the surface.

capacity of the bluegrass begins to decline, we have a second deep rich growth of alfalfa, brome grass and ladino in the rotated pasture lots. When the first field is moderately pastured off, we close the gate of the first pasture and open that of the second, returning to the first only to mow down whatever vegetation has been left. When the second rotation pasture is pastured down, we open the gate of the third lot and mow off the second. By the time they have pastured down the third lot, the first field is carrying from a foot to eighteen inches of luscious new growth of brome grass, alfalfa and ladino. The procedure is repeated through the summer until the animals are taken into the barns for the winter.

This process of rotation has increased our pasture carrying capacity by at least a third and greatly increased our milk production per head since the dairy herd is always on green fresh growth whether on bluegrass or on mixed legumes and grasses. If a season should arrive when the bluegrass plus the rotated pastures failed to provide sufficient pasture, there is, under our grass legume program, always plenty of forage and we need only open the gate of any meadow and turn the herd in upon a second or third cutting of rich brome grass, alfalfa and ladino.

Throughout the summer the dairy herd runs on the 28 acres of bluegrass where there is abundant spring water and abundant shade. Moreover, the combination of bluegrass bottom and heavy, rotated brome grass-alfalfa-ladino pasture provides a variety of choice and mixture of grasses, legumes and weeds which contribute not only to milk production but to the good physical condition of the cattle as well, for they are able to obtain in the weeds along the ditches, the creek and the wet spots a number of deep-rooted plants for which they show a great taste, doubtless because they are finding in them a number of minerals and trace elements which are not present in the shallow-rooted bluegrass and white clover and also because certain weeds will concentrate in their leaves and stems certain minerals which the cattle crave. Under careful observation we have noticed that cattle, even on the finest bluegrass and white clover or alfalfa, brome grass and ladino, will eat yellow dock, mint, young nettles and many other kinds of weeds. I am willing to trust their taste and judgment since cattle know better which plants and minerals are good for them and know better how to balance their own diets than any of us, from college professor to veterinarian.

In the case of our heavy meadows of legumes and grasses, the fertilizer and man-hour costs are so low that we are able to maintain a twenty-four-hour feeding schedule during the winter with no fixed feeding hours. This permits the animals, including the milking herd, to eat as and when they please as they do on pasture. It permits each cow to balance her own diet of dry hay, minerals and grass-legume silage to suit her own taste, always a good procedure, and with our good grass silage we reproduce very closely the conditions of summer pasture. The cows do nothing but eat, lie down, make milk, get up when they are hungry, eat, lie down and make milk again. The system has increased our winter milk flow by 10 to 15 per cent.

In fact, under our grass farming program there is very little difference in the milk production of our herd between May and June and August and September or January and February. In summer they are always on good lush pasture because there is an abundance of fields in brome grass, ladino and alfalfa which can be used and shifted about on a flexible program where at any season it can be turned into grass silage or hay or pasture, and in winter with grass-legume silage and good hay, fed twenty-four hours a day, the conditions of good summer pasture are virtually reproduced in the loafing-shed.

This maintenance of uniform milk flow is economically important for two reasons: (1) When prices go down in the flush spring season we are making a good profit and when in shortage time in the hot summer months or in winter the prices rise steeply, our milk flow shows very little variation. (2) Our *minimum production* base, upon which good spring prices for milk are based, is very close to our maximum production, varying little more than a can a day the year round. And all this without protein supplement or concentrate, with minimum grain feeding and with an abundance of the finest roughage it is possible to grow, produced at a very low cost of labor and fertilizer. There is no "summer pasture" problem at Malabar and no necessity for raising Sudan grass or other makeshift substitutes for top-quality pasture.

These are the reasons why we claim to make from 30 per cent upward more net profit than the average dairy operation. They are important factors in the economics of a dairy farm or indeed any livestock farm. Our operations are not based upon a hope of high prices but upon making a good profit by efficient and economical operation regardless of price. When the price of milk falls we are doing all right and when it is high we are making a great deal of money, on a basis of maximum production per acre of maximum-quality feed at minimum costs in fertilizer, labor and supplements. I am assuming of course that one also keeps good, high-producing cows and not "boarders," and maintains a proper breeding cycle.

At the time of writing we are contemplating using the excess first cutting as bedding. This first cutting is very bulky and contains great quantities of rather coarse brome grass. A good part of it is put green into the silos and the remainder could well be used as bedding since it is produced cheaply and in great quantity with no plowing, fitting and with only small amounts of fertilizer. In other words, like salt marsh grass, it "just grows" and it is very likely that it costs us less to produce than the straw from small grain crops which require the labor of plowing, fitting and the expense of fertilizer. We should then of course purchase the barley and oats needed to put in the grass silage, probably at a cost little higher than that of raising it. This would be a streamlined grass farming program at virtually the maximum.

One of the most profitable factors in our dairy-livestock program is the use of Balbo rye for late fall and early spring pasture. In the area around the main dairy barn it becomes necessary every four or five years to renovate the rotated brome grass-alfalfa-ladino pastures. This is not because the seeding has gradually "worn out" but because, after four or five years of trampling by the cattle, the earth becomes impacted and, although the sod is still abundant and heavy, the impaction limits the moisture-holding factor, the aeration of the soil and consequently the growth and bulk production per acre of the grass and legumes. In order to maintain maximum production, these fields are periodically torn up and reseeded.

No mouldboard plow goes into the field, but in autumn the Graham-Hoehme plow is brought in and the field is ripped from end to end once over and then again at right angles to a depth of a foot or more. In this process any hard pans are broken and much of the surface trash and the roots are thoroughly worked in to a considerable depth. Once over with the Ferguson offset disk (one of our most valuable tools) and the field is leveled for drilling rye. The soil is left loose and open with the consistency of truck garden soil and the means of a deep, effective drainage is established which is important because the rye will be pastured in the late winter and early spring months when the fields are wet.

This operation takes place in late August and early September so that the rye will have a heavy growth by the end of October. The rye is cross-drilled at the rate of 4 bushels to the acre, not only to provide a maximum of heavy pasture but also to promote the maximum

[177]

of root growth as organic material which helps mightily to counteract the impaction caused by trampling.

By the end of October there is a heavy pasture at least a foot in height which in a good season has been known to carry the milking herd through until Christmas. Immediately after snow flies and the ground is frozen, the field is given a heavy coating of good barnyard manure which not only fertilizes the field but adds greatly to the organic content so important in reducing compaction later on. Winter snow and rains wash out the "strongness" of the manure and by March the rye is again nearly a foot deep. Long before the bluegrass is ready for pasturing the rye provides us with heavy green forage. After the mild winter of 1949 we turned on to rye pasture on March twenty-ninth.

When the bluegrass is ready for pasturing, the cattle are taken off the rye, and again the field is torn up with the Graham plow and the Ferguson disk and is seeded to spring barley or oats, according to the type of soil in the field in which a seeding of brome grass, alfalfa, orchard grass and ladino is made. (Oats are seeded on the heavier soils and barley on the lighter ones.) In late June or early July, between cuttings of hay, we harvest the crop of oats or barley, keep the grain for next year's silage and take off a considerable quantity of straw. By the late autumn after the average frost date we are again pasturing the field, which after an average summer has a rich growth of ladino plus the alfalfa and grasses. Thus we have taken two valuable crops in a year off the same field while actually *improving* its fertility and greatly increasing the organic content, not to mention the fall pasture on the new seedings.

In an average season, the practice gives us out-of-doors green pasture at least two or three weeks longer in the fall and two or three weeks earlier in the spring than can be provided by any of our regular pastures. The program keeps milk flow at maximum, permits the animals to act as their own manure spreaders and saves much labor in indoor feeding. It also makes for great economies in hay and silage feeding at both ends of the pasture season and makes unnecessary the extra labor in summer of putting up winter forage for an extra four to six weeks of the winter season. It is important to realize that it is not necessary to pasture all day on the rye. The green rye acts as a strong stimulant to milk production and, as a stimulant, an hour or two a day of pasturing is sufficient to increase greatly the milk flow even though the cows are being fed only dry hay.

Such an all-year program keeps the dairy cows on fresh green

[178]

pasture for most of the year. Beginning in March they go on to the rye. Two to three weeks later there is a rich growth of bluegrass ready for them which carries them well into June when the abundant rotation pastures of brome, alfalfa and ladino are ready after the first cutting has been taken off the grass silage. During the summer the cattle run between the bluegrass, white clover "tramping ground," with its abundant shade and water, and the rotated fields so that, in effect, they have everything that should make a cow happy and induce her to produce as much milk as possible. Under the usual autumn rains, the bluegrass begins again to produce heavily and provides plenty of pasture in case an early or heavy frost cuts down the ladino-alfalfa production. When the bluegrass is pastured off or ceases to grow as the weather becomes colder, the Balbo rye is standing a foot deep waiting to be utilized. Balbo rye is used because it is a rye bred for pasture or cover crops to be turned in, and produces twice the leafage or more of ordinary rye as well as a heavy and organically valuable root system.

During the coming season we anticipate one more step in the streamlined approach to dairy and livestock farming. During the two months or more of the hot, dry months we shall harvest fresh legume and grass forage in the meadows and bring it to the dairy herd at least once a day, allowing them at the same time the freedom of roaming if they choose through the bluegrass bottoms where there is abundant shade and spring water. Under this system, the cows will have the maximum amount of top-quality feed throughout the hot months without the effort of finding it for themselves and so will avoid the drop in milk flow which comes when cows find it too hot to stir from the shade. As every farmer knows, the more good forage a cow consumes and the more good spring water she drinks, the more she will produce. By this method we shall also increase very considerably our carrying capacity per acre, even over the rotated pasture system, since all trampling wastage will be eliminated, and under the regular cutting system, with the use of a forage harvester-field chopper, the production of tonnage per acre is very greatly increased. The forage, of course, will always be of top-quality, succulent character, with no tough or withered grass or legumes whatever.

Under this system we anticipate conservatively a 20 per cent increase in carrying capacity and at least a 10 per cent increase in milk flow. Again this is along the lines of maximum production per acre on the land we already own rather than extending ourselves on to new land without increasing our profits per acre. It is all in the di-

[179]

rection of the kind of intensive farming in which lie the *big* profits of the farmer. The cost of the extra labor, we anticipate, will be paid about four times during the short period of the hot months by increased milk production and increased carrying capacity per acre. The plan would be especially profitable for dairy farmers on smaller acreages from 120 acres upward, both in increased forage yields and increased milk production.

We also anticipate the feeding of weaned lambs under a similar program in pens and without grain feeding, simply turning the raw material of our lush grass and legumes into meat, carrying the lambs from late June to December first.

These things are important in livestock management for in the case of milk, and even to a degree in the case of feeder animals, the farmer cannot look for steadily rising profits from higher and higher prices. One of two things happens. Too high prices restrict the market, the number of buyers declines and the price falls again, or eventually the government will tell the farmer how much he will get for his milk and his meat. The increased profits can and must be found not at the top but at the bottom of the production market, in more efficient planning and management and better forage and feed produced at low labor and fertilizer costs through grass and legume programs and higher feed yields per acre. In many cases, by some reshuffling of program and operations, the dairy farmer's net profits on the same herd and the same amount of land could be increased all the way from 25 to 75 per cent and occasionally even more.[3]

[3] In some of the most famous milk-producing areas of the nation I have seen farmers trying to make a profit with herds running on wretched, burnt-out pastures throughout most of the summer months. In order to keep up the milk flow, these same farmers are spending hundreds and even thousands of dollars in feeding hay and grain and expensive supplements when they could produce more milk and have healthier animals by spending a small fraction of the sum once every three or four years on pasture improvement to grow their own forage and protein. It is usually this category of farmer who finds that milk prices are never high enough and who is always asking for subsidies and support prices while the answer not only to his security but his prosperity is directly under his feet. Moreover, higher prices for meat and dairy products means in turn higher wages for industrial workers and higher prices for everything the farmer buys. As his prices rise, the purchasing power of his dollar shrinks since that is the way inflation really works despite the illusion shared by many indifferent farmers who expect prices rather than good management to make them prosperous. The only possible way in which inflation helps the farmer is in paying off money borrowed on a note or mortgage in terms of a sound dollar with the money represented by a cheap or inflated dollar, which is of course a poor basis on which to practice agriculture. The fact is, as pointed out earlier, that what is needed in agriculture and in the nation as a whole is not more and more dollars which constantly buy less and less for everybody but solid dollars which buy more and more. The farmer

Following a period of experiment by ourselves and by Ed Babcock at Sunny Gables Farm we are embarking in 1950 upon one more step in streamlined production, reducing labor costs, achieving a better feeding program and increasing profits. The new step consists in making a whole feeding ration in the silo itself by adding grain to the silage as it is made. No corn will be used but only oats and barley (not ground) shoveled into the grass legume mixture as it goes through the silage cutter or is blown into the silo in case a field-chopper-harvester is used.

The process operates in several beneficial ways: (1) The grain, during the fermentation process taking place in the silo, turns to brewers' or distillers' grains (one of the finest livestock feeds) but with no nutrients removed. (2) The starch element changes largely to sugar and alcohol acting as a silage preservative and predigesting the grain for the cow's stomach at the same time. (The stomachs of cattle were not designed by God and Nature to digest grain and, as every farmer who has fed dry grain ground or otherwise well knows, 40 per cent upward of dry grain, fed separately from roughage to the cattle, goes through them without being digested and without the animal's deriving any nutrition whatever from it.) (3) In feeding the predigested brewers' grain coming out of the silage, *all* the nutritive value of the grain is utilized by the animals. The fact that the grain is actually *mixed with* the roughage serves also to allow the cow's stomach to act upon it as if it were really a part of the roughage and therefore to increase greatly the utilization. In short, we have found that one could cut the grain ration almost in half and still achieve better results in milk production and the cow's general condition over feeding dry, ground grain separately in the old wasteful traditional fashion. Indeed, Mr. Babcock found that one had to watch the amount of grain mixed with the silage as, even when fed in comparatively small amounts to heifers, the animals tended to become too fat to breed well or milk well after freshening. After considerable experiment we have come to the conclusion that whole oats and barley in the amount of approximately 100 pounds to the ton of silage is about the right mixture for heifers and dry cows, 150 to 200 pounds for cows on the milking string and 300 to 400 pounds for fattening out feeder cattle.

A good deal of experiment has been done recently with considerable

who constantly expects his prices to save him from his own laziness, ignorance or lack of efficiency is a doomed man and will prove to be more and more so in the coming years.

success in feeding grass-legume silage to hogs, and the same system of feeding grain and even ground corn (cob and all) could be applied to hog feeding by increasing the amounts of grain per ton of silage, thus cutting the feeding labor virtually to the supplying of minerals and meat scraps or tankage at free choice to the hogs. Again this is a streamlined operation for it produces quickly a good meat hog bringing top market prices with a reduction in the cost of labor and feed.[4]

The process of making a whole ration for the cattle right in the silo itself cuts by 30 or 40 per cent the wastage of expensive grain and achieves better results in growth and production. It guarantees excellent silage and it cuts the winter feeding labor by more than 50 per cent, eliminating all the carting of grains to the hammer mill, the grinding and the feeding it back again to the cows. In other words, we do the whole of our winter grain feeding during the ten or twelve hours we spend on a summer day filling the silo. On a farm such as Malabar, producing dairy cattle and milk, there is no need for corn whatever as the combination of oats and barley makes a much better feed than corn for dairy animals.

It should not be overlooked that barley or oats are cheaper to raise than corn if one is talking feed rather than cash crops. You plant them and you harvest them with a fertilizer cost a good deal less than that of corn. In our own case small grains fit into our program in a way which makes the total profit per acre probably equal or more than that of corn, since we utilize the straw from the barley and oats for manure which we produce in great quantity. Considering the fact that we cannot fertilize the heavy meadows more often than once in three years because of lodging and harvesting troubles, our manure goes farther each year as fertilizer and our bill for purchased commercial

---

[4] Even in the heart of the corn-hog area in Iowa, farmers have been moving more and more in the direction of raising their spring litters on grass-legume mixtures, notably ladino, alfalfa and brome grass, with an absolute minimum of corn feeding. The process provides a quick-growing lean hog bringing top market price, and the less corn fed the hogs the greater the profit since, whether the farmer buys corn at a high price or whether he raises it at high costs of labor, fertilizer and gasoline, corn is the most expensive feed a farmer can employ. As one farmer put it, "I used to feed about 12 bushels or more of corn to put a hog on the market. Today I feed at most 3 or 4 bushels. Figure in for yourself the price of corn, whether I raise it or buy it, and see the enormous difference in my net profits made by using grass and legumes as the chief feed." Here again is a way for the farmer to reduce greatly his feeding and labor costs and at the same time produce a top-market-price animal or commodity by using brains rather than muscle and gasoline. Of course, the immense value to the farmer's soil of a program which shifts the emphasis from soil-destroying corn to soil-building grass and legumes is one of the most important factors and benefits to the farmer himself.

fertilizer steadily grows less. If you ask where the mineral fertility is coming from to make up the lag, that factor is explained in detail in the chapter called "Farming from Three to Twenty Feet Down." It is coming very largely from our deep minerally rich subsoils in organic form as heavy sods and as the excess minerals which the animals cannot assimilate and for which they have no need, since in the deep-rooted legumes and grass the mineral nutrition actually exists in an excess amount. This excess in minerals goes back to the fields in the form of barnyard manure.[5]

It is necessary for us to produce straw since we cannot buy it in sufficient amounts without shipping it from considerable distances at excessive costs and, through regard for our soils and organic material and the general cleanliness of our animals, we prefer straw to using sawdust or shavings. If straw were easily and cheaply available we should probably increase our profits by not raising even oats and barley but by buying what we need and putting the acreage of barley and oats to grass and legumes and carrying more animals.

After streamlining crop production, it became evident that we needed a streamlined livestock program which would give us the greatest benefits and profits on the basis of grass, legume and small-grain farming. This did not form itself at once but came about through general evolution in which many things had to be considered, not the least of which was our limited cattle shelter and feeding barns and their old-fashioned construction. They were conceived and built in the days when labor and time were cheap, during a form of agriculture much closer to the pioneer-frontier school than to our modern streamlined plans. As with the average farmer, the changes in program have been made slowly and conservatively, many of them with our own labor and materials; but by replanning the big and high, old-fashioned bank barns, we were able to cut labor and man-hours by at least 50 per cent. The biggest barn was converted into a

[5] It was once commonly and universally believed that all barnyard manure was alike and contained the same fertilizer values. We know now that barnyard manure can vary as much as vegetables in its mineral content according to the kind of soil from which it comes originally in the form of forage. In other words, poor land and poor farms with soils of depleted or unbalanced mineral content produce poor manure and rich soils produce rich manures because the cattle cannot utilize all the minerals coming to them through the forage off rich soils. This factor of course has little to do with the "inoculating" values of manure which carries benevolent bacteria, fungi, moulds, animal secretions, hormones, enzymes and other elements which undoubtedly affect the germination of seeds and set to work reactions within the soils which make the mineral content more available to succeeding crops. Nor has it, of course, much to do with the value of the organic material represented by the manure which is very great.

[183]

combination loafing-shed and milk-parlor with hay and straw storage directly overhead so that both could be fed directly to the cattle through trapdoors or open feeding racks extending to the second floor. This arrangement meant merely cutting the strings on the bales of hay or straw and allowing gravity to carry them downwards. The other barns were remodeled in a similar fashion save that there was no milk-parlor.

The cattle program went through many changes from the original combination of beef and dairy cattle until gradually the beef cattle were eliminated entirely and under the grass-legume-small-grain program we came to have dairy cattle only.

At the present time Malabar brings in carloads of Holstein heifers bought in Wisconsin, Minnesota or Canada when the prices are low in the late autumn. These are turned in at breeding age with Angus bulls. We take from them a calf and a year's milk and resell them as second-calf heifers, for two and a half times upward what we paid for them, carrying them meanwhile on excellent, heavy, cheaply produced pasture in summer and in winter on good, cheaply produced grass and legume silage and hay, without grain save when they are on the milking string. There are many advantages to such a program and many profits, arising mostly from the money we do not spend in nitrogen fertilizers, in feed supplements, dairy rations, grain feeding and man-hours of labor.

The operation is a comparatively simple one with a high degree of flexibility. When we have a heifer which is wild or a hard milker or a low producer, we do not bother with her in the dairy, but turn her out on good grass as a nurse cow to two calves. At the end of the year she delivers to us two calves which have cost us, directly, zero in protein supplements and grain, zero in calf feeds, zero in bucket feeding or labor of any kind and zero in sickness. In some cases the profits on the two calves are higher than if we had taken the heifer into the milk-parlor. The heifers are bred to Angus bulls for two reasons: (1) Most of them calve eventually on pasture and the cross-bred calf has the small, compact shape of an Angus calf so that calving difficulties are minimized and very rarely have we had any difficulties. Each day the herd of dry cows and heifers is watched and new calves are picked up, put into the back of a jeep and fetched home with the mothers following. (2) The cross-bred calves are all polled, nine out of ten wholly black and uniform in appearance, a factor which makes them desirable as feeder calves to prospective buyers. They also make quick-growing, profitable feeders with the

[184]

usual cross-bred vigor. If we have more calves than nurse cows to care for them, the calves are sold as veal, and at the end of the season, about November first, all calves from week-olds to 400- and 500-pounders are sold off and the stage cleared for the new heifers coming in from Wisconsin, Minnesota or Canada. Calves born during the winter are doubled up with nurse heifers and given a creep where they can learn to eat silage and hay and when spring comes are turned out with the nurse heifers on pasture.

When we have an exceptionally fine producer, we do not sell her but keep her for our milking herd, thus building up a top-quality producing herd much more rapidly than we should be able to breed such a herd. The breeding and resale of heifers at two and a half times what we paid for them, plus a calf and a year's milk or two nurse calves per heifer, represents a rapid turnover of capital, and the calves on each November first, when sold, represent a cash crop fully as much as a crop of wheat or corn and they are considerably more profitable in the short as well as the long run of soil building and steadily increasing fertility. Meanwhile a fat milk check keeps everything going.

This program, based upon streamlined, high-quality, maximum-production, low-cost forage and farming has by no means been pushed to its full limits. We hope within the next two years to go into the cheese business which will about triple the gross returns on our milk, and we shall have in addition 2000 to 3000 pounds of whey each day. This brings us again into the hog business, not to the corn-hog business, but on a basis suitable to our kind of hill country and much more profitable than the ordinary corn-hog operation. Pigs can be grown and fattened quickly on heavy alfalfa and ladino and whey during the summer months with a minimum of grain and in winter upon whey and grain-reinforced grass-legume silage, again with a minimum of grain feeding. In other words, the hogs will not be fed on high-cost corn grown at high labor and fertilizer costs but as a by-product of grass-legume farming in which, under lower costs and as a by-product, they become very nearly all profit.

It might be enlightening to survey the year's work program at Malabar under the existing streamlined program. Beginning with March when the season opens, we put in enough oats and spring barley to provide us with sufficient straw and a surplus of grain. The fertilizer cost is much lower than that necessary to obtain a good crop of corn. Once the grain is planted and the field seeded to brome grass, alfalfa and ladino, there is no more work on those fields until we harvest the grain and bale the straw. In April, May and June there is no plowing for corn, no fitting of the field several times, no planting and no three-time cultivation, nor any of the heavy fertilizer costs of corn. Late in May the heavy grass-legume meadows are cut for silage. By the time this operation is finished, the oats and barley are ready for harvesting and the straw for bailing. When the grain and straw have been harvested, a second cutting of grass and legumes is ready to be made into high-quality hay. We make as much as is needed for the winter and pasture the rest of the second cutting. We also pasture the third cutting or, on ground in need of improvement, let it stand. Then from the middle of August we have nothing to do in the way of field work except to plow at our leisure, any time from then until Christmas, the legume sods we will plant to barley and oats the following spring. We never have to plow hard ground in order to get in wheat in a rush. We can wait for the rains. We have no corn to pick or harvest or put into the silo. We plow when we feel like it in a leisurely way when the conditions are right and then permit the fields to lie over for spring oats and barley planting.

Even in this operation we cut man-hours and gasoline costs, for the rough-plowed land is worked upon by frost action which kills the uprooted alfalfa plants and other vegetation and fits the land and mellows the soil far better than any machinery can accomplish the same job. On our gravel-loam soils, one fitting with a culti-mulcher in the spring is usually sufficient to level the ground for the grain drill to work over. Moreover, instead of turning under a raw new sod which will make heavy demands upon the nitrogen and steal it, together with moisture from the growing crop, the sod is already dead, decaying and contributing nitrogen and organic fertilizer rather than stealing it, and providing a high degree of tilth favorable to the following crop in terms of viability and of moisture conservation. Very rough plowing in the fall prevents any real run-off or erosion even on our hills and every bit of melting snow and rainfall is absorbed by the earth. Three times during the year we haul the rich manure out of the loafing-sheds and spread it over the fields.

Since the abandonment at Malabar of the mouldboard plow, gasoline and soil-fitting costs have been more than cut in half. The Graham-Hoehme plow is used to rip up the sods during the autumn months, going over the fields twice, the second time at right angles or diagonally. By the second time a maximum depth of fifteen to sixteen inches has been obtained, breaking up any old hard pans and creating no new ones. All moisture penetrates to that depth where it is conserved for the following season, and the top rough surface is left so well drained that it is possible to go in and level and fit the field for planting at least two or three weeks earlier in the spring than we could go into the same field to plow up a heavy sod (an important factor to high production in the case of oats and spring barley). The spring fitting is usually a once-over operation since the soil is mellow and filled with dead and decaying organic material. Moreover, the Graham-Hoehme plow, in going once round a field, does the work at a greater rate of speed, so that the man-hours and gasoline employed in plowing a field once over are cut to approximately a fourth of the time and gasoline needed for the same operation done with the mouldboard plows. All this is also in line with our belief at Malabar that the less fitting done, the better for the soil, and that the more we keep tractors and machinery off the fields, the greater are the gains for our soil and for our production per acre.

In 1950 we anticipate carrying the "streamlining" one step further by a plan which at first glance might appear to be revolutionary in the extreme. The idea first came from Bob Huge, my partner in Texas and the manager at Malabar, Ohio, and when I first heard it, I confess to a shock which may approximate that of the reader, but when a pencil and paper were brought into use (a process neglected by too many farmers who frequently raise crops at a loss without knowing it), the new plan became not only reasonable but actually profitable. It was simply to use first-cutting hay rather than straw as bedding for the cattle at Malabar.

As explained earlier, one of our problems has been to get enough straw for bedding without putting an unprofitable quantity of our land into small grains. We consume each year not only the straw we produce but some of the neighbor's straw as well and at times still run short by spring. (All of this is, of course, converted into that most valuable of fertilizers, barnyard manure.) Gradually we have reduced small-grain acreage until we reached that point where we simply ran short of straw for bedding long before the end of the winter. All the time growing right under our noses was the solution in profitable terms. It was to eliminate all row and grain crops, go whole-hog for grass and legumes, and use our excess or spoiled hay for bedding, buying what small grain we need.

The first cutting of the alfalfa, brome, ladino meadows is truly prodigious, up to four tons and better of dry hay per acre in which

[187]

the brome attains a height of four feet and the alfalfa three. A good part of the first cutting goes to fill seven and, in 1950, nine silos for winter forage, but there still remains each year a heavy tonnage and a big acreage of this first cutting hay which in good weather is difficult to cure properly and in thundery weather, in late May and early June, virtually impossible to cure into top-quality hay. In the past the worst of this hay was actually used for bedding and for mulch on fruit trees and vegetable gardens where it was of great value because of its freedom from all ripened weed seeds which might germinate and make a nuisance later on. From 1950 on we plan to use *all* of the excess first cutting for bedding which, since cutting out corn and small grains, will provide us with an abundance of bedding at low cost.

I can understand at once the reader's reaction that we should sell this hay for which, when well cured, we can get between $15 to $20 a ton, and at first glance such a judgment would seem reasonable. But again nothing in farming is simple, and to arrive at a proper balance of costs and profits it is necessary to consider many other factors which are set forth below.

Against the price per acre we might get by selling this hay (and not all of it is of top quality owing to poor curing conditions) we must write off the man-hours, the gasoline, the wear and tear on machinery and the heavier fertilizer charges which arise *annually* in the raising of even small grains. (The meadows remain a minimum of seven years without putting a plow or a fitting tool or a drill in the field.) The grass and legumes simply grow year after year with no man-hours and no gasoline save in harvesting and with 200 pounds of commercial fertilizer per acre or a top dressing of manure every third year. Moreover, by putting *all* our land into grass-legume mixture, the carrying capacity of cattle on the farm both winter and summer is increased by nearly 30 per cent on a regime in which good grass-legume hay, silage and pasture are the backbone with no protein-supplement feeding and a minimum of grain. This in turn greatly increases the supply of manure and organic materials to the benefit of all our soil and steady increase in production and in its value on a basis of capital gain.

In turn the process permits us to sell such expensive machinery as a combine, grain drill and virtually all our plows and harrows, for under such a regime we shall run a big livestock farm with a Graham plow, an offset disk, a hay bailer or field chopper and a silage cutter in case we continue to use the bailer for making silage. This means again another cut of close to $5000 in expensive machinery on which we no longer need to charge off interest and depreciation at the rate of several hundred dollars a year. Actually we can buy the small amount of corn, barley and oats needed in a grass-legume dairy program much cheaper than we can raise it.

In addition to all of this, our man-hours in the field are again immensely cut leaving hundreds of man-hours a year for painting, concrete work, better care of the livestock and the upkeep which is so valuable in the daily profits as well as in the capital value of any farm. And, above all, we have the satisfaction of seeing our land and particularly our hillier, rougher land growing steadily greener and richer and more productive through the use of grass and legumes. It should not be overlooked that the organic and fertilizer values of barnyard manure made of grass and legumes are much higher than those made from a base of straw, sawdust or shavings.

To many readers the Unimal theory of Ed Babcock is familiar. If it is not, I suggest that they write him for information regarding it. Mr. Babcock's theory is simply that too great a proportion of our population is ill fed in terms of what he calls the high-protein or "refrigerator" foods—meat, dairy and poultry products, milk, butter, etc. This is so because these things cost too much as they are now raised and it is this same high cost which restricts markets and creates surpluses, most of all in the field of corn and wheat. Babcock believes that we should have many more meat and dairy animals as well as poultry, produced and fed on an efficient and profitable basis, which could consume these grain surpluses and lower the prices of "refrigerator" food. This is of course the New Agriculture at its best and of course it implies an agriculture on a much higher level than any we have been accustomed to in the past. Our own plan of going over wholly to grass and legumes and buying from the grain farmer what grains we need is an integral and valuable part of such a plan. Under it I challenge any comer with the assertion that we can, under efficient farming and livestock operations on such a base, produce meat and milk and dairy animals at a net profit at least 30 per cent greater than the conventional four- or five-year rotation general farmer raising his own grain and using corn silage. The 30 per cent is a very conservative statement. The net profit, in reality, is perhaps 50 per cent upward rather than 30 per cent. Moreover, we have on our land absolutely no erosion, soil fertility gains rather than depletion, and less than 5 per cent water loss under any known conditions. Certainly the system makes most corn farming look unprofitable and even silly unless the farmer is raising from 100 bushels an acre upward, an accomplishment limited to less than 3 per cent of our corn farmers.

Of course such a program reaches toward one of the fundamentals of a sound and prosperous agriculture—that of proper land use which in essence means simply the proper consideration for soils and their evident adaptability to certain specific crops and the growing of the right crops on the right land and in the right area. We can on our glaciated hills produce better quality legume and grass forage, both in quantity and quality, than some of the richest Iowa corn land, and it is suicide to raise corn up and down hill in our area. To get reasonable yields of corn without heavy losses of soil and water it is necessary in our country or on any rough land to grow it on stripped, contoured or terraced land, a procedure which is unnecessary on rough or hilly land to grow grass and legumes and even in some areas small grains. Therefore let those of us on good hilly land grow the maximum in quantity and quality of grass, legumes and pasture and the farmer on flat, rich corn land grow the corn and the grains which we will buy from him for supplementing roughage feeding. Both will profit and such a practice would do much toward reducing the surpluses of grain which become at times extremely troublesome in an economic sense.

Proper land use, of course, stands as the fundamental cure to many of our agricultural and economic problems. It is something which we have as yet by no means achieved in a national sense. In this pattern, of course, there is small place for the old-fashioned, pioneer type of general farm with its highly dispersed and distracted work plan and its necessity for excessive investments in a wide variety of tools and machinery. Proper land use does in many senses represent specialized farming but it also represents efficient and profitable farming and the possibility of adapting whole farms and indeed whole areas and building the soils of such farms and areas to the level of maximum production of one or two given crops instead of soils which are in essence a compromise, adapted to *no one crop*, and unsuited in one degree or another to *all crops*. In other words it is the extreme soils problem, stated in an earlier chapter, of the farmer who tries to grow both alfalfa and potatoes; one or the other will suffer and in no case will he achieve maximum profitable production of both. At Malabar we are in one of the finest grass countries in the world and in grass and legumes lies our best hope of survival and prosperity and high profits. Even under the finest conditions our light soils do not produce the high yields in corn or small grains, especially in any dry season, of heavier and leveler soils, but in grasses, legumes and pastures it is possible to raise high yields of the finest quality. That is the fundamental reason we have moved more and more toward a grass-legume-roughage program with its elimination of expensive machinery, high fertilizer and fuel costs, high man-hours, and high costs of supplements. Under such a program we can undoubtedly buy any corn or small grains far more cheaply than we can raise them, and at the same time benefit the growers of corn and small grains in areas better adapted than ours to their culture.

I am aware that, superficially, such a program might not meet with favor from the manufacturers of fertilizer and farm machinery, but I think that such a reaction would be both superficial and wrong. The prosperity of the farmer is the best assurance of prosperity for the manufacturer and indeed the prosperity of the whole nation. The farmer who makes plenty of money can buy all the fertilizer he needs and will not skimp on it but only increase his prosperity by buying and using it wisely. The farmer with plenty of money will buy the best machinery and buy it more often and he will have more money to buy radios and automobiles and books and a hundred other things not only because he has money but because he has, under a wise program, time to enjoy all these things. And he will have as well the pride and dignity that always goes with a sound agriculture. He will be working many fewer hours and making much more money than most of us farmers have had in the past, and the drudgery will have largely gone out of farming in the dignity of a profession in which brains and not brute force is the measuring stick of the good farmer.

I am aware that the conditions which made possible a streamlined program at Malabar are by no means universal. The pattern however is applicable in one way or another to most of the vast Middle West and the whole of the South, and it may contain hints for similar methods in other areas where efficiency and better regard for the soil and for grazing can increase our wealth and lead to an easier life for all of us—a life not dependent upon government hand-outs but upon a sound and a profitable agriculture. It is a program particularly suited to millions of acres of potentially rich and productive rolling land which has been badly treated and partly depleted by a wretched row-crop, cash-crop agriculture, and it opens up the prospects of profits per acre as great as those to be had on much of our high, flat, production corn lands—or probably, when all the costs are in, much greater profits.

[191]

CHAPTER X

Go South, Young Man!

*I attended a funeral once in Pickens County in my state
[Georgia]. . . . This funeral was peculiarly sad. It was a
poor "Gallus" fellow whose breeches struck him under his
armpits and hit him at the other end about the knee.
They buried him in the midst of a marble quarry. They
cut through solid marble to make his grave and yet the
little tombstone they put above him was from Vermont.
They buried him in the heart of a pine forest, yet his pine
coffin was imported from Cincinnati. They buried him
within touch of an iron mine, yet the nails for his coffin
and the iron in the shovel that dug his grave were imported
from Pittsburgh. They buried him by the side of the best
sheep-grazing country on earth and yet the wool in the
coffin bands and the coffin bands themselves came from
the North. The South didn't furnish a thing on earth for
that funeral but the corpse and the hole in the ground;
there they put him away and the clods rattled down on
his coffin. And they buried him in a New York coat and
a pair of breeches from Chicago and a shirt from Cincinnati, leaving him nothing to carry into the next world with
him to remind him of the country in which he lived and
for which he fought for four years but the chill of blood
in his veins and the marrow in his bones.*

—HENRY GRADY

(Note. *Henry Grady was one of the great Southerners
and perhaps the first of the New Pioneers who understood,
earlier than any Deep Southerner, the true reasons for
the terrible decline into poverty of the Old South, and the
enormous evil done to soil, people and the nation by the
one-crop, tenant, sharecropper agriculture ruled by King
Cotton. It may be of interest to learn that in one Southern
town there is a monument erected to the boll weevil,
because the depredations of this insect drove the farmers
out of the poverty of cotton agriculture into the prosperity
of balanced farming.*)

X. Go South, Young Man!

IN THE course of a year a great many young men and women drive into Malabar Farm to look around. Most of them are about to start out in life and many of them have very little money and few possessions outside of the clothes on their backs. Usually they are intelligent, energetic and ambitious. Some have college educations and a good many have a deep interest in agriculture. They are this generation's crop of young men and women ready and aching to make their own way and in doing so to build up the nation as a whole. Until a couple of generations ago they could have gone out and claimed a section of land or a tract of forest or gone prospecting for fabulous mineral wealth. All they needed to start farming was a team, a plow and a harrow. That was the time when Horace Greeley said, "Go West, young man!" Today there isn't any more virgin land or any more wealth to be had for the taking. The Old Frontier has gone forever, but a New Frontier has taken its place and that New Frontier is, paradoxically, in one of the oldest parts of the nation—the Old South.

Usually after we have talked awhile, I finish up by saying, "If you want opportunities, go South! The whole of the South is waking up. There are opportunities there which no longer exist in the overcrowded North and Northeast or even in the Great Plains or the mountains where everything long since has been taken up. In the Deep South there is a whole country in the process of being made over. Why go to Alaska or Brazil when there is a pioneering job right at home under your very nose?"

For a long time now a revolution has been going on almost unnoticed below the Mason-Dixon Line. Only those with the seeing eye and comprehending mind have noticed or understood what was in progress. It is a revolution compounded of many elements—the gradual disappearance of the old professional Southerner who looked for an excuse for his own failings of character and energy in the War Between the States and in the freeing of the slaves; the fortunate and gradual disappearance of the "honey-talking," professionally Southern females who pretended not to know where babies came

[195]

from but could give many a doxie lessons in how to land a man; the mechanization of agriculture; the rising influx of industry, the progress of education; the great improvement of agricultural methods and proper land use; the co-operative, intelligent and constructive power of local and state banking methods; and the increasing number of energetic young men and women with a sense of responsibility toward community, state and nation.

All of these things and many more are contributing mightily to the revolution which is making the South a wholly new world of wealth and opportunity.

First of all, it would be well to identify what the writer means by the South. Secondly, it would be well to state a few simple home truths. It seems to me that the *real* South includes Virginia, North and South Carolina, Georgia, Alabama, Mississippi, Tennessee and East Texas. It cannot be said, I think, that Florida, with its resort attractions and the general Coney Island atmosphere, could be called the South. Louisiana, with its Gallic code of laws and behavior and its peculiar politics is only partly typical. It is always Louisiana, a state apart. Central and West Texas and Oklahoma are really the Southwest from which the First Frontier, on which a man could build a fortune out of a two-bit piece, has not wholly disappeared. Arkansas is border country which has never seemed able to make up its mind, and Kentucky and West Virginia are definitely border states. Missouri, despite the pretensions of some sentimental inhabitants, never was and is not today the real South.

The first unpleasant truth is that the South would very likely have reached the depths of permanent economic depression even if there had never been a Civil War. This is so because, in the world since 1800, any purely agricultural nation or area has been at a great economic disadvantage. This in turn, is so because, with the passing of cottage industries during the Industrial Revolution, virtually the whole of income of such areas or nations is drained away in purchasing nails, machinery, clothing, kerosene lanterns, coffins, machine tools, and what you will, elsewhere.

Not only, however, did the South suffer from the economic hardship of a purely agricultural society, but the situation was made worse and more destructive by the fact that it suffered as well from one of the worst agricultures in the history of the world. It was an agriculture supported by so-called "cheap" slave labor which grew constantly more expensive as agricultural income declined, and it was a careless one-crop "cotton" and in small areas one-crop

[196]

"tobacco" agriculture which steadily depleted the soil and permitted millions of tons of rich topsoil to be washed away each year. It was essentially an agriculture of depletion which created constantly declining yields per acre and constantly augmented costs of production. The Civil War in reality contributed only incidentally to the economic decline of the South. It merely hastened by a couple of generations what was already inevitable.

Of this blight by poor agriculture and the need for supplementary industry I can think of no better illustration than that of a trip over the back roads through the Deep South made by the writer many years ago. After driving for miles through worn-out and eroded cotton lands without seeing any dwelling better than a wretched cabin or a crumbling, deserted plantation house, we would suddenly pass a fair-sized and habitable house painted and in good condition. With a smile of pride, my driver would invariably turn and say, "Coca-Cola money!"

In other areas along the Atlantic seacoast where Northerners had bought up and restored old half-abandoned plantations, a different driver would say at the sight of a well-kept, handsome house or a prosperous farm, "Yankee Money!" That Yankee money was, of course, really Southern money which should have stayed there but had gone North and come back again.

Over great areas of the Deep South today the blight of a wretched agriculture still remains. In other considerable areas hundreds of thousands of miles of terracing have been created to prevent soil erosion and the depletion of the bare cotton fields by winter rains, and here and there in the worn-out, blighted areas one finds communities, areas and individual plantations where the landlords are prosperous, the houses are painted and new tractors and automobiles stand in the plantation yards. In relation to the New Frontier these prosperous establishments are like the original great plantations built long ago in the savannas and the virgin forests of the First Frontier. They are built in a new wilderness created by man out of the original, greedy, careless one-crop agriculture which largely ruined the South. One was the symbol of man's conquest of wild, undisciplined Nature; the other—the prosperous modern farm or plantation—is the symbol of man's conquest over his own carelessness, ignorance and greed.

But the New Frontier is not merely an agricultural one, although that represents perhaps the fundamental solution of the Southern problem. It is a frontier where there are new opportunities in business, small and large, industry, processing, servicing and many other fields. Part of the remarkable and increasing revival of the South is based upon a better agriculture, producing more per acre of all crops at a steadily declining production cost and a resulting increase in profits, purchasing power and cash resources, all of which affect the whole economy of the area in terms of employment, savings, bank deposits, purchasing power and in countless other ways.

But equally important have been the increase and dispersal of industrial enterprises which have provided not only employment but steady, year-round employment rather than the short, poorly paid and inadequate seasonal employment provided by a cotton and tobacco agriculture. This has contributed and is contributing more and more to keeping money in the South rather than exporting it northward to workers who once manufactured very nearly everything the Southerner bought but his food. That balance between industry and agriculture—a balance between a good agriculture with its accompanying and important purchasing power and the local manufacture of commodities and employment which provide the soundest economic base of any area or nation—is beginning to be a reality over an increasingly large part of the South.

This balance has a bearing on many of the economic problems of the South. One of these problems—that of higher shipping rates for freight into and out of the Deep South—has been a source of continual Congressional wrangling and dispute for nearly two generations and is at present the subject of a suit brought by the state of Georgia still awaiting hearing before the Supreme Court. The "discriminatory" rates have been a definite handicap to the development of the South and in particular of the more remote areas.

The variance in Southern freight rates from the rest of the country, however discriminatory, has arisen in the past from sound economic reasons. It is somewhat like the question of which came first—the hen or the egg. Few, if any, railroads show a profit for passenger traffic alone. Their profits come from freight carriage. During the long poverty of the Deep South, freight carriage was light and in a one-crop cotton area very often largely seasonal. The South produced only cotton, tobacco and some timber to be shipped, and in most areas cotton production per acre has been declining steadily since before the Civil War, so that the railroads carried steadily less and less. In the exact ratio to the declining yields, purchasing power decreased and the amount of goods shipped for sale *into* most areas constantly declined. The passenger traffic, owing to the poverty of the Southern areas, was at the same time both light and cheap.[1]

These conditions have long been recognized by the government agencies which granted the railroads operating in the Southern areas higher rates on freight shipments than those operating in other parts of the United States and in particular in the Northeast, where both freight and passenger traffic were and still are so great as actually to tax at times the capacity of the carriers.

The achievement of a better agriculture-industrial balance plus a better and more productive agriculture and livestock enterprise will greatly increase need for shipping and very possibly may eventually correct the "discriminatory" shipping rates on a basis of economics alone, regardless of any action by government agencies or Supreme Court decisions.

But all of this is changing and in many areas far more rapidly than many Americans or even Southerners realize. The spread of industry through the South has increased at a phenomenal rate in the last ten years. Partly this has occurred through the wise policy of government during the war in dispersing new war plants throughout the country rather than increasing the concentration of them in the already over-crowded industrial Northeast. Some of the increase has come about through the wise realization of many great corporations, and of working men themselves, that the increasingly crowded cities mean poorer and poorer working conditions, higher living costs and a declining standard of living. Part of it has come about through the magnet to manufacturers of cheaper wages. This advantage will, as industry spreads through the South, largely disappear and has, indeed, already begun to do so.

Whatever the reasons, the money of the South is beginning to stick where it belongs instead of being drained off. The improvement on the agricultural side, which I proposed to discuss a little

---

[1] In some parts of the Southwest and in areas once famous for high agricultural production there exist today "ghost" towns created by a steadily declining agriculture. In areas where the old-timers report that yields of cotton, wheat and corn were once so heavy that these commodities were stacked along the railroad track awaiting shipment, there are towns in which the population has almost vanished and the stores and banks along the main streets are closed and boarded up. In some areas of such former abundance railroad tracks have actually been pulled up and the countryside abandoned save for a few miserably poor inhabitants. Single-crop farming and erosion have done their work in such areas sometimes within the short period of less than two generations. Of course, the economy neither of an area nor of a nation can long withstand such a process.

further on, is of equal or even greater importance in bringing about the same result.

Among all the Southern states and, indeed, even among the whole of the forty-eight, no state has shown so much progress within the past generation or less than North Carolina. It is a progress which is not economic alone but possesses countless social and cultural facets as well. Few states have made so rapid an industrial advance and no state, in the South at least, has advanced so rapidly toward the achievement of the vital industrial-agricultural balance so important to the balanced economy and stabilized prosperity of any modern state or region. It is not so long ago that in passing through North Carolina the same poverty and shabbiness which has largely character-ized the Old South since the Civil War was almost everywhere in evidence. Today the shabby look has largely gone. The farms look prosperous and well cared for in most areas, and the shabby cabin slums on the fringes of the towns are very definitely on their way out. In many areas they are gone altogether. The University of North Carolina at Chapel Hill has become one of the most impressive cultural centers of the nation.

It is, I think, indisputably true that culture, social advance and civilization in general are to a large extent tied to economics, to living standards and to productivity. So are ignorance, prejudice and in-tolerance. Education plays its role, beyond question, but education, too, is tied to economics and prosperity or at least to a stable, fairly prosperous and balanced economy. Here again North Carolina is proving the theory, for as a state she has made remarkable progress away from the old superstitions and the prejudices which in the past handicapped and at times paralyzed the Deep South.

It is not my purpose to attempt to prove these things. The existence in the past of ignorance, prejudice and intolerance is largely admitted most frankly by progressive Southerners themselves. These things had their roots very largely in the poverty of the South since the War Between the States, and as that poverty wanes and is succeeded by economic prosperity, many of the old troubles correct themselves, as indeed they have already begun to do in North Carolina and in those areas of the Deep South where a better agriculture and sound indus-trial growth plus better incomes and living standards have advanced side by side.

The bitterest racial feeling in the South has never existed between the Negro and the most prosperous elements of Southern society but between the handicapped Negro and the unfortunate and poverty-stricken elements of the white race, who found in the different color of their skins their only dubious claim to superiority and in the Negro the scapegoat to be blamed for their own failures and miseries.

The so-called "liberal" elements have long railed at Rankin and the late Bilbo, called for their impeachment and even conducted political campaigns against them within their own state of Mississippi, but the fault and the cure lies *not* in the impeachment of either or both men but with the low economic level of the state of Mississippi. One could do away with Bilbo and Rankin but the chances are that men very like them would soon bob up again, for the simple reason that under the representative form of government, they represent Mississippi and cannot be done away with until Mississippi itself is cured, and cured by strong doses of economic prosperity which in turn brings better living standards, better education, tolerance, and abates the need for scapegoats and hatreds arising largely on poverty alone. A better agriculture, a more dispersed industry and a balance between the two can bring about this prosperity with its attendant benefits and in many areas has already done so. In the Deep South, North Carolina stands today at one end of the scale and Mississippi at the other. The reasons are largely economic.

Despite the influx of industry, the economic backbone of the South is still largely agricultural and the gains in this field have been important and are becoming increasingly so. The agricultural income of North Carolina has increased in five years from about $200,000,000 to about $800,000,000. Even if we should write off 40 per cent of this gain as arising from inflated prices, the gain would still be remarkable. It has been achieved largely through better and more balanced agriculture, through increases in technology, mechanization and most of all through soil conservation, good land use and the gradual abandonment or reform of the old vicious system of single-crop tobacco and cotton agriculture.

I have a deep affection for the South and an enthusiastic interest in its present and future, particularly in the fields of agriculture and forestry, and periodically I make a tour along back roads, through swamps and small towns and pine woods, and each time I revisit the same territory I am astonished by the progress in prosperity and living conditions and in happiness. There are certain farms and plantations I like to revisit again and again because there is nothing pleasanter than visiting people who each year become happier and more prosperous, and to watch the change upward in a whole countryside and landscapes.

I like to revisit my friend Perkins in Alabama. He is now in his sixties. He came home from the First World War to go to work as a tenant farmer with a team of mules, a plow and a harrow. Last week the local banker told me that Perkins was close-mouthed about how much he was worth but that he owned 3000 acres, hundreds of cattle and was worth pretty close to $1,000,000.

And there are the Simpson Brothers in the same state who about fifteen years ago inherited 600 acres of cotton land with a $40,000 mortgage which was probably more than the whole place was worth. In those days the whole plantation was in cotton year after year with a per acre yield of less than half a bale of cotton. Today the boys are raising more cotton on 50 acres than the whole of the farm raised when they took it over. The rest of the land is given over to sweet potatoes, pecans, grass and legumes, beef cattle and hogs, with occasional seed crops. And the 600 acres has grown to 3000 acres. Because their families outgrew the original plantation house, the brothers have built two handsome new brick houses in the pecan orchard which would rouse pride in the heart of any big business executive. In fields heavy with green pasture in the month of February, there were hundreds of fat cattle, and down by the river bulldozers were at work reclearing land which had once been cleared by the original pioneers, worn out, abandoned and left to second-growth timber. It is good land, but ignorant and greedy farming methods had reduced its yields to a point where it was no longer worth farming. These new pioneers, the Simpson Brothers, will make lots of money out of that so-called "worn-out" land where the bulldozers were busily working.

And there's my friend Tom Necker, a little shriveled fellow of about sixty, over in East Texas. He hasn't a lot of land and until ten years ago he had it all in cotton. Then on a government loan he bought some cows and planted some grass and legumes. The gross return on his land the last year he raised cotton was less than $1000. Last year his gross on the same acreage was over $10,000, and as his land continues to mount in fertility under a grass, small-grain dairying program, the gross will increase, even if prices drop.

And there is Cason Calloway, a rich man, who at forty retired from the cotton manufacturing business to give the rest of his life and energies to Southern agriculture. He chose the "poorest" land he could find in Georgia and went to work on it. To say that today it "blossoms like a rose" is an understatement. He has used his money, together with some contributed by other Georgia businessmen, to help set up pilot farms with energetic young men to work them in the prospect of becoming proprietors themselves in a few years.

And Mack Gowdy, who grew up under the handicaps of the once devastated South and by his own brains and initiative has made himself a little kingdom where absolute security reigns and crop failures or poor yields are unknown.

There are scores and hundreds and presently thousands of cases like these, many of which I know, many of which I have never seen, but their achievements and their prosperity is making itself felt for great good not only in the economy of the South but in that of the whole of the nation.

You well might ask, "How did these men achieve such things out of the wreckage they inherited from the Old South?"

The answer is complex. The rebirth has come about through the brains and energy of individuals like those above who have shown the way. It has come through government financing and the devoted work of ill-paid county agents and soil conservation and T.V.A. agricultural engineers, in some cases through the wisdom and foresight of local country bankers like Bill Bailey in Tennessee and Will Campbell across the river in Arkansas, of big influential bankers like Robert Hanes and the Wachovia Bank in North Carolina. It has come about through the contributions of dozens of farmers, of experiment station and agricultural college professors in finding new crops and new uses for the good red and black soils of the South. And below all of this lies the fundamental fact that the great proportion of the soils react richly to knowledge, science and good treatment, and the fact that today we know how to create soils which are better and more productive than most virgin soils laid down by Nature herself.

Limestone has been one of the great factors in the revival of Southern agriculture, for it acts as a catalyst to make other minerals available to plants, animals and people and it makes possible the growth of the miracle-working legumes which more than any other factor have brought back prosperity to the agricultural South. And the legumes feed the whole range of succulent grasses which in turn make for profitable beef and dairy cattle production. In some areas even lime did not make possible the growth of alfalfa or wholly promote the growth of other legumes. Then agricultural science found out the missing factor. It was one of the trace elements—boron. When two 5-pound packages of borax per acre were applied to the soil, the alfalfa and other legumes flourished.

The chain of limestone, legumes and grasses has increased the organic content of the soil which in the past, under single-crop cotton, had been reduced to zero. The result is very nearly a miracle.

A vast part in the story has been played by new crops and newly developed seeds and agricultural products. Again the contributions have come from every sort of agency and individual. Sometimes, as in the case of a grass known as Kentucky Fescue No. 31 and in that of the Cayly or Singletary pea, the valuable crop was growing in the South all the time under the very noses of the farmers and experiment stations, simply waiting to be discovered. Kentucky Fescue 31 has been growing almost since the beginning of time in the Kentucky-Tennessee area, but only a few years ago a county agent, in his trips around the countryside, noticed a steep and rough hillside which was deep green in midwinter with growing grass. The next summer he gathered seed and the agricultural agencies began propagating the grass. Now the seed is available to any farmer, and over thousands of acres in the South the grass is providing succulent pasture out of doors for livestock right through the winter months.

The Cayly or Singletary pea has been growing in fencerows and along roadsides in the Deep South as long as the Kentucky Fescue has been growing in Kentucky, but it took a farmer called Cayly, who had come down from Ohio to Alabama, to observe that this wild plant grew and remained green and flourished throughout the winter months. He saved some seed and began propagating it. Cayly's observation has been worth millions to the South for it has provided a legume that grows all through the winter and, after being pastured, can be plowed in as organic material along with the great quantities of nitrogen which it has fixed in the soil out of the air. The legume shares the name both of Cayly, the Ohio farmer, and of Professor Singletary of Louisiana State Agricultural College, who made the same observation as Cayly at about the same time. The names are used interchangeably in the South, but the legume is the same. What matters more than the name is the wealth its development has brought into the South.

Cayly, incidentally, is another friend who took over "worn-out" land on the edge of the once fabulously rich black belt of Alabama, built it back, and has become a rich man. He came down originally from Ohio, planning to farm in the South in winter and the North in summer, but the extraordinary advantages of climate and the long seasons and the potentially good soil in the South changed his plan and he farms there the year round, returning each summer only for a visit to his Ohio farm. There may well be a profound symbolism, affecting the whole of the livestock industry, in this one farmer's migration from the Middle West into the South.

Many of the plants which have played an enormous role in the rehabilitation of the South have come from remote parts of the world and found Southern soils and climate singularly adapted to their growth. Alfalfa, the "green gold" of agriculture, is credited with an Asiatic origin, yet Darwin in his *Voyage of the Beagle* in 1820 reports great patches of a beautiful green legume plant flourishing in the arid desert of Chile. He was told that it was called "Alfalifa." Whatever its origin, it has become one of the great wealth-producing crops not only of the South but of the nation, and not only because of its high protein and feeding values but because it provides a sound check upon erosion and is a great producer of nitrogen deep in the soils. More and more since the definite requirements for its growth have become known it is becoming a "poor land" crop used to provide a profitable crop on "worn-out" soils while it "builds" the soil as well.

Kudzu comes from Japan. It is a deep-rooted leguminous vine which makes a prodigious growth each summer, sometimes as much as forty to fifty feet. It has existed in this country since clipper-ship days when it was brought in and used as an ornamental vine on porches and was commonly known as "porch vine." Someone discovered its fabulous qualities in covering gullies too deep to be bull-dozed over. It not only provides an absolute check on deep gullies and steep banks but furnishes first-class, high-protein pasture and forage. It might be called with equal justice the "last-resort" vine, for it is efficacious in blocking erosion where all other methods fail. Because of its capacity for prodigious growth, it is regarded by some as a nuisance when it becomes uncontrolled and runs wild. Throughout the South it has reclaimed land once regarded as hopelessly lost, and in Georgia its admirers, under the leadership of Channing Cope of Atlanta, have even organized a state "Kudzu Club" to sing its praises and advance its use.[2]

Crimson clover, one of the great crops of the South and, like the Austrian pea, one of the best of winter cover crops and nitrogen-producing legumes, comes from Europe and Asia. Both grow prodigiously in the Southern climate and in properly managed soils, creating nitrogen and organic material and cheap, high-quality forage for livestock.

[2] Channing Cope is the author of an excellent and entertaining book on Southern agriculture called *Front Porch Farmer*.

All plants have the capacity for occasionally producing "sports" or variations which sometimes are of great value, and such a sport occurred with crimson clover a few years ago in Alabama. In a pecan orchard, seeded to crimson clover, a patch about ten feet square appeared which reseeded itself not immediately, as was the habit of this soft-seeded plant, but a year later, the seed remaining dormant in the ground during that time. Gradually it took over the whole of the orchard. Its great value immediately became evident in the South, for such a seeding habit made it possible to grow another crop such as corn or cotton during the intervening time and, when that crop was harvested, to find the same field covered with a thick carpet of young crimson clover plants which had seeded themselves. This fitted perfectly into the needs of the South where one of the great problems had been to find a "cover" crop which would keep the earth protected, following row crops like corn and cotton, against the erosion caused by torrential winter rains. This "sport" of crimson clover fills the need and without reseeding. It reseeds itself if permitted to form seed at a year's interval, provides cover, fine pasture, nitrogen and organic material (one of the great needs of Southern agriculture), all, as you might say, painlessly and without trouble. I saw fields of this clover which had been seeded five to six years earlier still flourishing without any reseeding, although the fields had grown fine crops of cotton and corn in alternate years. This valuable sport of crimson clover is now known as Autauga County Reseeding Crimson Clover after the county in which the variation occurred in a farmer's pecan orchard.

The lespedezas, a whole family of forage-producing, high-protein legumes, come from Korea and Manchuria and flourish better in the South than in their native background. Ladino clover, a close rival of alfalfa as "green gold," was brought to this country about twenty-five years ago from Northern Italy. Under proper rainfall conditions it is the greatest producer of high-protein pasture for cattle, hogs and poultry now known. It, too, covers the ground and builds rather than destroys soils. It is a factor in the enormous growth of the chicken and turkey industry which has added millions of dollars yearly to the income of the Southern states.

Another foreigner of Asiatic origin is the blue lupin, a legume which can be sown in the autumn after the harvesting of corn or cotton or peanuts to cover and protect the earth against winter rains. It germinates quickly, grows through the winter, produces blossoms in February and is ready to plow under with all the nitrogen and organic material it produces in time to secure another bumper crop of corn or cotton the following season. Sometimes it is used for grazing before finally being turned into the soil.

All of these innovations and importations have raised the agriculture of the good farmer in the South far above that of the King Cotton days when year after year cotton was grown without putting back anything into a soil which was left bare to be washed away each year by the torrential winter rains. One might almost say that these grasses and legumes are very nearly worth their weight in gold, for their use is translated into dollars and cents at an increasing rate each year. Moreover, their use means farming less acres to produce more cotton or corn or beef or milk, with the steadily declining costs of labor, seed, taxes and interest and the rising margin of profit which such a process brings about.

Strawberries, asparagus, truck crops, peanuts and other diversified products have added more and more millions to the agricultural income of a vast area where once cotton was the only crop and grew up to the front door of every cabin and plantation house. The growth and sale of tomato plants, seeded early in the South and shipped North for replanting by Northern tomato growers, has become a whole industry, and tung oil, a valuable ingredient of paints and varnishes and indispensable to some of these, is being produced by an increasing number of orchards given over to the tung tree, one of the most beautiful of trees. The tung tree comes from China and formerly the whole of our tung-oil supply came from China. In 1947 the Gulf states produced 66,700 tons at $100 a ton.

One of the most dramatic stories of plant breeding in relation to the South is the story of the yam, the sweet potato and Dr. Julian Miller of Louisiana State Agricultural College. The yam and the sweet potato have long been Southern crops, but their great economic value as a source of wealth is only beginning to be realized.

A number of factors—better cotton yields per acre (leaving more land free for other crops), grasses, legumes, livestock and in fact almost everything but the ubiquitous, highly subsidized cotton—have contributed to the story, but one of the greatest contributions has been made by Dr. Miller who worked to develop out of old and frequently deficient strains afflicted with runtiness, stringiness, bad keeping qualities, etc., a kind of super yam and sweet potato. It was a long battle with years of work and many disappointments, as only those familiar with plant breeding can understand, but Dr. Miller finally achieved his purpose and his super yam has come to replace thousands of acres of cotton and to swell the income of those same acres as much as two or three times. In the wake of the super yam have come canning factories and dehydrating plants and more employment and more money flowing in from the North instead of being drained off from the South wholly into the Northern industrial areas.

Much of the vast new wealth of the South has come out of the rapidly growing cattle industry, for it has at last been realized that the Southern livestock grower and dairyman has great advantages over farmers engaged in the same enterprises in the North and the Middle West. Through most of the South the great capital investment in the huge and expensive barns and equipment to house livestock during the long winter is unnecessary. At most a rough shelter against cold rains is needed and in some areas and with beef cattle even such rudimentary buildings are wholly unnecessary. In a great part of the South it is possible to keep livestock on pasture throughout the whole of the winter. This means little or no labor costs either in winter feeding or in spending the whole of the summer months to harvest and store forage for cattle during the long winter months. The money and labor which the Southern farmer is *not* forced to spend represents an enormous economic gain and profit. Moreover, the livestock man with his program of grasses and legumes and manure is constantly building the fertility of his soils and the capital value of his land instead of tearing them down.

Montgomery County, Alabama, once the very heart of a great cotton-growing country, today derives the major part of its income from beef cattle. Second comes dairy products and third cotton.

Until very recently most of the milk consumed in the South came from the North and the Middle West and from as far away as Minnesota and Wisconsin. Everywhere milk was expensive—too expensive for the budgets of many families—and in some smaller communities it was simply unobtainable. Even today many cities from Houston and Dallas to Atlanta and Winston-Salem import a considerable percentage of the milk consumed from the Northern tier of states. The situation, however, is rapidly changing, and it is not inconceivable that in a few years the South may well be an exporter rather than an importer both of beef and of dairy products.[3]

[3] Owing to the simultaneous operation of two Malabar Farms, on the same dairy-cattle, milk-producing program, one in Ohio and one at Wichita Falls, Texas, we have been able to make very apt and accurate comparisons. The operation in Texas represents at least 60 per cent less capital investment in buildings and shelter and 45 per cent less expense in operations and at the same time we receive from $1 to $1.50 more per 100 pounds for milk.

Everywhere in the South today one sees fields which only a little while ago were bare and gullied now covered with grass and legumes in which fine herds of beef and dairy cattle wander about up to their knees in fine forage.

One of the soundest developments in the South has been the gradual disappearance of the old plantation, absentee landlords and the passing of the land into the hands of men working the land as owners rather than under the miserable tenant-farmer, sharecropper arrangement, which in the past contributed so much to the poverty, misery and low living standards of the South. It is simply a truism that unless both owner and tenant are enlightened, both steal as much as they can out of the soil, with the result that the economy of a whole area or the whole of the nation suffers as well as promoting the steady decline in income of both tenant and absentee owner. No other single factor has destroyed more good land in the nation than the absentee-tenant, short-term, year-by-year arrangement. The change for the better in the South is of great benefit to everyone in the nation. In 1930 some 40 per cent of the farms in the South were farmed by owners and 60 per cent under the absentee-tenant arrangement. In 1945 58 per cent were farmed by owners and only 42 per cent by sharecroppers or tenants. Since 1945 the percentages have continued to move in the proper direction. This means not only a better agriculture but better diet, better living standards and the kind of individual security which is indispensable to sound democracy.

Statistics regarding increased agricultural income in the South are nothing less than formidable—showing an increase from about $2,500,000,000 in 1940 to nearly $8,000,000,000 in 1947. Even allowing for inflated prices and for the general rise in national farm income during that period, the figures as in the individual case of North Carolina, are remarkable. Whereas the gross farm income in the whole of the United States increased over the same period 106 per cent, the gross income of fourteen Southern states increased 148 per cent.

In the past most eroded, abandoned, "worn-out" land was considered only fit to return to forests or wilderness. The South, more than any region, has shown and is showing that this is neither necessary nor true. In the past the South, more than any other region of the nation, has suffered from erosion, depletion of organic material and a general decline in yields per acre of all agricultural commodities. While cotton has been perhaps the largest single contributing factor, other elements beyond the direct control of man have done their part in the destruction, such elements as heavy seasonal rains and high temperatures and mild winters which contribute to the burning up of organic materials. However, until very recently, Southern agriculture took few steps to combat these disadvantages. Rather it aggravated them by a single-crop agriculture which left the earth bare to erosion the year round, returned little or no organic material to the soils and constantly stirred them by open-row cultivation, thus increasing the rate of burning-up of organic stuff.

There are more miles of terracing in Georgia and Alabama today than in any other state of the union, but the terracing (to prevent erosion) came about largely through desperation. It was a question of checking the erosion or of abandoning the fields altogether to scrub pine and wilderness. We know now that terracing is by no means the final answer to the destruction of agricultural land but only the first step which, unless followed by a better agriculture and land use, becomes meaningless and in some cases can even be destructive. The real answer lies in rotated crops, the restoration of organic materials through green or barnyard manures and the use of grass and legumes.

As has been written many times in this book, millions of acres of so called "worn-out" soils both in the South and in the rest of the country are not worn out at all; they have merely been farmed so badly and so ignorantly that the native mineral fertility, often very great, has simply become unavailable to the plants, animals and people through lack of lime, of organic materials, of moisture, of bacteria and of many other elements necessary to a living, productive soil and a sound and profitable agriculture. This is particularly true of the South.

We know now that most of the soils of the South, however eroded or even unproductive they may be, are fundamentally good soils and can be restored to a high degree of fertility at an expense which is by no means prohibitive. This holds true in most of the red soils and certainly of the eroded, marl-based soils of the Black Belt of Alabama and Eastern Mississippi. It is certainly true of the almost inexhaustibly rich alluvial soils of the great Delta region. The problem is of building up those soils and of making available to plants, through moisture, organic material and other means, the native fertility of the soils and subsoils and even of the commercial fertilizer used on those lands.

Hundreds and even thousands of Southern farms established on "worn-out" cotton lands are showing, through diversification, livestock, and legumes, that this can be done, and as they prove the point, the income, economic status, human dignity and purchasing power of the farmer and the other residents of any given region rise magically. They are proving a point of utmost economic value not only to the South but to the nation—that millions of acres of low-production or even abandoned land can be transformed from a tragic economic and social liability into a great national asset.

It is a simple rule applying alike to industry and to agriculture that the more one produces per unit, per man-hour, per dollar invested in taxes and interest, the lower the cost of production and the wider the margin of potential profit. The South, agriculturally speaking, has been a perfect example of this rule in the past, since before the Civil War until very recently the production per acre of cotton declined steadily. Many farms which on virgin soils produced as much as 2000 pounds of cotton per acre, reached a low point at which they were producing as little as 200 to 300 pounds per acre. The result was of course disastrous since the cost of plowing, fitting, planting, cultivation and harvesting of the cotton remained the same per acre or even increased while the production per acre steadily declined. The same disastrous declining ratio held true as to taxes and interest while land values themselves declined in some cases to nothing at all. In the same areas where disaster gradually overtook the agricultural economy, the graph of production per acre has begun to mount sharply as production costs have declined at an equal rate. There are farms which today are producing by an improved agriculture as much cotton on 10 acres under cultivation as they produced on 50 or 60 acres only a few years ago. The remaining 40 or 50 acres have been put into profitable use in raising livestock, sweet potatoes or other crops.

The dethroning of King Cotton or at least his reduction to the status of a constitutional monarch is perhaps the most fortunate single economic happening in the history of the South. No tyrant king ever did more damage to a people.

Of the income dollar of Southern agriculture in 1925, 53 cents came from cotton. In 1946 only 20 cents came from cotton. The important fact is that the number of income dollars was more than three times greater and that most of the increase in the number of dollars came from the dethronement of King Cotton and the increase in grass, livestock, poultry and other non-cotton production.

At the same time, the purchasing power of agriculture has been going up by leaps and bounds. The number of tractors in use on Southern farms shows an increase of 155 per cent as compared to an average increase of about 100 per cent elsewhere. Today the electrified farms in the South represent 40 per cent of the nation's total. These and countless other statistics of similar significance are revolutionary and represent tremendous gains which affect not only the South but the economy of the whole nation, for their effect is felt wherever industry manufactures commodities for sale.

While agricultural prosperity is inevitably based first upon the fertility of the soils and the yields per acre, technology and mechanization have in the case of such a product as cotton produced truly great benefits, cutting costs of both production and processing and increasing margins of profit.

In West Texas they are producing cotton on soils and under rainfall conditions which are adverse in comparison to most of the Deep South at a greater profit than many a cotton farm in the rich Delta area of Mississippi and the reason lies in mechanization and technology. In the past cotton was always grown the hard way, indeed the hardest way possible, by slow hand labor, on soils which became each year more difficult to work and less productive, with an endless amount of hauling and handling. The field was plowed and fitted by mule power, the excess cotton plants "chopped out" by hand, the rows weeded and hoed by hand, the cotton picked by hand. But the process did not end there. The cotton was then hauled miles to an old-fashioned inefficient gin, then frequently enough back again to the plantation and finally perhaps miles in another direction to the "compress" which baled it. It was small wonder that all this, coupled with low production per acre, made it impossible to grow cotton profitably and saddled all of us with millions of dollars a year in taxes to support the inefficient and backward cotton agriculture and industry.

Today it is possible to "chop" cotton by machinery, weed it by flame throwers, pick it by pickers or strippers and finally to gin, clean and load it all in one spot within a few minutes. Moreover, instead of picking cotton two or three times a season by hand as it ripened, a "storm-proof" strain has been developed which remains on the stalk so that it may be harvested in one picking, mechanically, after the plants have been defoliated by a chemical spray. In one county in West Texas in 1948, there were 1800 mechanical strippers at work. These mechanical developments plus a good agriculture and a high production per acre at low cost should make any support price or subsidy of cotton wholly unnecessary and, indeed, make it possible for the American cotton growers not only to compete with but actually undersell the cheap labor of Egypt, India or China.

It might be suggested that such technological and mechanical developments would cause great unemployment, but that is where spreading industry can take up the slack and is doing so. Actually in West Texas and in some other Southern and Southwestern areas, a high degree of mechanization has been brought about largely by the increasing cost of field labor through scarcity or total disappearance of hand labor from fields into factories where it receives the higher wages of industry. Again this development has tended to provide all-year-round employment instead of merely seasonal employment and at higher wages than in the past, thus raising living standards and purchasing power and creating a new economic dignity for the individual worker.

The sum total of a poor cotton agriculture and poor yields coupled with archaic methods of cultivation and processing have made cotton, generally speaking, less an economic asset than a serious liability and a burden to the nation as a whole. The future of cotton, save on a basis of high production per acre plus improved technological methods, seems dim indeed, and a good many farmers in the South would be wise to think of other crops in the future and in particular of livestock, especially since, with the advances of rayon and other synthetic materials, the demand for cotton is constantly decreasing and will continue to do so unless new uses are found for this agricultural commodity. In the meanwhile, the program of the government of buying and storing cotton to keep the prices high has, together with the high cost of production under an inefficient agriculture, virtually priced cotton out of the foreign market. Moreover, the government subsidy program which, in effect, is what the government buying program really is, serves to maintain in operation hundreds and even thousands of inefficient and bad farmers, without initiative, who, save for the support buying, would go into some other crop or leave their land for steady employment in some industry and permit it to be farmed by men who understood and practiced a sound agriculture. The government buying program cannot go on forever, especially as the 132 million non-farming taxpayers of the nation are growing steadily more aware of the costs of such a program and are becoming increasingly impatient with it. Surpluses of single-crop corn and wheat are serious enough, but at least they represent food. Surpluses of a non-edible commodity in a declining market seem much more inexcusable, especially since there is nothing on earth to prevent many cotton farmers from shifting to a more

[213]

productive and profitable farm program, except perhaps that the shift might represent more work

The so-called cotton bloc of Southern Democratic congressmen, operating under the log-rolling system of "I'll vote for your bill if you'll vote for mine," has been able to maintain this paralyzing subsidy long past the period when it served its purpose as an economic cushion to tide the South over a period of wretched and bankrupt agriculture into a period of reformed and profitable agriculture. Actually the subsidy has served to retard rather than advance the economic progress of the cotton areas of the South.

About three years ago, the support price measure came within a narrow margin of being voted out and only a last-minute summoning of Southern Senators and a delayed vote saved the subsidy and perpetuated this needless and, indeed, corrupting burden on the taxpayers of the whole of the nation. Until it is abolished, cotton agriculture will never stand wholly or honestly on its own nor show the enterprise and improvement which can greatly benefit the South by eliminating the bad absentee landowner and bad individual farmer. It serves to tie up much potentially good agricultural land which might otherwise become an economic asset and to preserve the medieval Southern system of sharecropping and tenantry, one of the prime causes of poverty as well as ignorance, prejudice and intolerance.

Despite these circumstances, the South is definitely on the way up and the New Frontier in agricultural land offers as great opportunities as the First Wilderness Frontier at less economic cost and infinitely fewer hardships than those known by earlier frontiersmen. I know of no better example than the southernmost area of the Tennessee Valley Authority near and surrounding Athens, Alabama. Driving from Birmingham to Athens one passes for miles through acres of worn-out land, badly farmed land with only here and there a passable good farm and decent buildings. As one begins to approach Athens there are signs of an economic improvement and prosperity which rises to a crescendo as one crosses the line into the T.V.A. area immediately surrounding Athens. Here and there are fields of legumes, livestock, good pastures and big yields. The farmers own shiny new automobiles and the houses look prosperous and well painted. Athens itself has long since lost the look of a run-down, shabby, Southern town. The difference within a few miles is very nearly unbelievable. It is merely the difference between good land use and bad, the difference between an agriculture of erosion and

[214]

depletion and that of legumes, livestock, lime, phosphorus and soil conservation established under the guidance and co-operation of the T.V.A.—the difference between the old, shiftless, greedy agriculture and the New Agriculture.

But the gains of the new South are not entirely economic and social. They are cultural as well. In the rising vitality of the literature in the South, one has again the evidence of the close relationship between the economic status of a region or a nation and its own contributions to civilization. The new literature out of the South is based neither upon the overheated romantic tradition employed as a justification for failure nor upon an inferiority complex justifying itself in excuses or in futile talk about the "damn Yankees." If it pictures poverty and ignorance and degeneration, it does not drape these things in pathos and romance but treats them satirically or in a fiercely condemnatory fashion. This fact is in itself of immense significance in any evaluation of the remarkable changes taking place in the South.

One of the tragedies of the whole great region is that the agricultural decline affected the people of considerable areas not only socially and economically but even physically as well. As Dr. Hugh Bennett of the Soil Conservation Service has so often said, "Poor land makes poor people," a fact which the medical profession is beginning to realize and understand to an increasing degree. The "poor white," the "redneck," the "peckerwood" are not alone the victims of the Civil War, of a declining economy and of an evil sharecropping tenant system. They are also the victims of the poor, monotonous and unbalanced diets induced by poverty and still more profoundly by the deficiencies of food grown upon originally poor soils or soils depleted by a poor agriculture which fail to make available to plants, animals and people the residual mineral elements absolutely vital in a sound nutrition. An improved agriculture, making these minerals available or replacing them where they are really lacking, means not only higher economic status, less ignorance, prejudice and intolerance but better physical specimens endowed with energy, active brains and inquiring minds.

Despite the great handicaps and the tragedies of the South, something has begun to stir mightily within the past decade or two. There is a whole new world opening up which manifests itself in a changed point of view, in a new liberalism born of Southern soil itself and represented by such men as ex-Governor Ellis Arnall of Georgia, Senator Lister Hill of Alabama, Representatives Estes

[215]

Kefauver and Albert Gore of Tennessee and many others both in state and national government.

It is not impossible that the solid Democratic South is on the way to breaking up just as the solid cotton South has already broken up.

This, too, would mark a great advance politically, as great an advance as the increase in sound agriculture and the influx of industry. For too many generations the South has found itself in the sad position of being the neglected child at the table. Because it has remained a one-party Democratic area, the Democratic Party leaders never needed to show the South any favors because they knew that their party would always carry every Southern state regardless of conditions, and the Republican leaders were, quite naturally, reluctant to show any favors to an area which persistently voted against their party. Only when both parties are forced to work for the vote in the Southern states will that area begin to receive from both parties the treatment and encouragement much needed to further the development of the area into perhaps the most important region in the United States.

Only as recently as ten years ago, President Roosevelt spoke of the South as "The Nation's economic problem number one." Before that time huge areas of the South were actually economic and social liabilities. The change has been remarkable and rapid but, what is more important, it has been sound, for it is no boom cycle based upon the discovery of oil or gold or the opening up of some new, virgin agricultural area. The hard-working, intelligent "new pioneers" of the South have taken over tired and worn-out land from the tyranny of King Cotton and turned it back into rich production. Cotton is allotted its proper place in a balanced agricultural program and where it is raised properly and efficiently it is a more profitable crop than it has ever been before in all its history.

CHAPTER XI

Poor Land Makes Poor Hunting

And so I shall proceed next to tell you, it is certain, that certain fields near Leominster, a town in Herefordshire, are observed to make sheep that graze upon them more fat than the next, and also to bear finer wool; that is to say, that in that year in which they feed in such a particular pasture, they shall yield finer wool than they did that year before they came to feed in it, and coarser again if they shall return to their former pasture; and again return to a finer wool, being fed in the fine wool ground. Which I tell you, that you may the better believe that I am certain, if I catch a trout in one meadow he shall be white and faint, and very likely to be lousy; and as certainly if I catch a trout in the next meadow, he shall be strong and red and lusty and much better meat; trust me, scholar. I have caught many a trout in a particular meadow, that the very shape and enamelled colour of him hath been such as hath joyed me to look on him: and I have then with much pleasure concluded with Solomon, "Everything is beautiful in his season."

—Izaak Walton, The Compleat Angler

XI. *Poor Land Makes Poor Hunting*

SOIL Conservation Chief Hugh H. Bennett is responsible for the saying that "Poor land makes poor people." He might well have added that "Poor land makes poor hunting and fishing" as well, for one will not find good fishing in eroded country where the streams are filled with silt nor in sandy country where the lake bottoms contain little vegetation and the water little or no plankton or the minuscule animal life which provide the major part of the diet of many varieties of fish. All kinds of game from the humble cottontail to the noble deer abhor worn-out and abandoned land where the soil has been depleted or the level of agriculture has become so low that the mineral fertility is no longer available to the vegetation and finally to the plants and the people living in such an area.

Animals, and perhaps even fish, know by instinct and by taste far better than people what is good for them. In most adults the instinct for good and minerally nourishing food has long since become diminished or completely atrophied. Many doctors assert, and much evidence backs them up, that small children still possess such an instinct and that the instinct should be used as a dietary guide as to what is good for a child. The instinct appears to lose its force sometime around the adolescent or Coca-Cola age, although it is not impossible that growing adolescent youngsters are right in their craving for doughnuts, ice cream sodas and other sweets; most of them between the ages of twelve and twenty have immense needs for the energy-producing foods both because they are growing rapidly and because they are immensely restless and active. Certainly most adults have lost all physical means of distinguishing by instinct or otherwise the nutritional content of vegetables or fruits on a basis of taste alone, although it is probably true that any vegetable or fruit with a high mineral and vitamin content actually tastes better than one grown on minerally deficient soil or under shortage of sunlight and moisture.

Science has countless examples of the power of animals to discern and avoid mineral deficiencies or to correct those mineral deficiencies by the most desperate means, if necessary. The sow who eats her young does so not through any depravity but because she has been kept penned up and improperly fed. She is hungering for protein as well as calcium and phosphorus. I have seen pregnant sows tearing up with the greatest effort and ingenious violence the concrete floors of their pens in order to chew up the cement and gain enough calcium to produce a healthy litter of pigs. A sow, set free with a new litter, will forage her way across country, taking care of them and bringing up a better, stronger litter than most sows penned and fed according to any but the best of dietary formulas. She will find somehow, even in depleted areas, the minerals she needs for herself and her litter. Cows will develop depraved appetites if fed from land depleted of phosphorus and calcium and take to chewing old bones in order to satisfy the deficiency. The average pregnant woman has lost such a faculty and minerals must be prescribed for her according to the knowledge of her needs available to doctors. She will not, except in rare cases, find those minerals for herself by the same desperate methods undertaken by the sow or the cow. I have even heard the fantastic tale of cows on the phosphorus-deficient Gulf Plain area of the United States eating dead fish washed up on the shore and even attempting to catch living fish in order to satisfy their craving for the deficient phosphorus.

And there is the famous experiment of Professor William Albrecht of Missouri University, one of the greatest authorities on soil and nutrition. In this experiment, stacks of hay harvested from five different fields, all showing serious mineral deficiencies in varying degrees, were set up in a winter feeding lot filled with cattle. The fifth stack was harvested from a well-managed fertile farm on which all the known vital minerals were present in the soil in good balance and available to the plants. For three successive years, regardless of how the stacks were arranged, the cattle, turned into the fields with the stacks, consumed the stacks in the exact order of deficiency of the farm soils on which the forage was grown, eating the nutritious stack first and then consuming the next best and so on down the line to the poorest.

And there is the now-famous story of the blind mule at Ohio State Agricultural College who was pensioned off and allowed to graze across the test plots of the college. Again one plot in this case had a well-balanced, productive soil, with five adjoining plots each arranged to demonstrate varying degrees of mineral deficiency. The blind old mule would stay on the fertile plot until he had grazed it down to the ground. Then he moved downward across the plots of declining fertilities in the exact pattern followed in Doctor Albrecht's experiment with the hay and cattle. Any farmer knows that if he fertilizes a strip or a part of a depleted pasture, the cattle will stay on the fertilized strip until it is eaten bare before going over to the pasture on the deficient, less well-balanced soil.

Curiously, wild animals, and perhaps to some degree fish, exhibit the same preferences for nutritious forage as against forage produced on poor land, or land so badly farmed that its fertility is unavailable.

The Ohio State Conservation Commission (a misnomer, for the Commission is concerned mainly with Fish and Wildlife propagation and habitat) has long made a study of the relation of game and fish to good or poor land and at the moment is compiling the findings and conducting final conclusive research on the subject. Simple observation and records have shown beyond any doubt the preference of wild game for fertile, abundant land and rich forage.

Ohio is, by comparison with many other states, one which has or had originally deep, rich soils. Much of the land has been depleted badly to the point that in spots where the woodchuck digs a hole, the subsoil he brings up acts like fertilizer on the depleted topsoil and produces rich and vigorous hay or small grain in fields where otherwise the crops show signs of the unavailability of minerals and deficiencies of all sorts. Oddly or perhaps naturally, wild life avoids depleted fields and areas and takes to the more fertile areas. In the violence of their distaste for poor land some of the cases approach the comic and create economic and even sociological problems.

The case of the humble cottontail rabbit provides a problem with which the game commission has struggled for years. The cottontail leaves the poor farm for the nearest good farm and he apparently leaves the depleted areas for the easy, rich living of the suburbs where gardens are fertilized and given abundant compost and other organic material with regularity. He has become a pest in city suburbs and in small towns while hunters over large areas complain that they cannot find in open wild country a single cottontail to shoot when the season opens. It is possible and probable that 50 per cent or more of Ohio's rabbit population resides and flourishes today *not* in the open fields of rural areas and second-growth timber but in the garden areas of suburbs and small towns.

For the past two years the commission has engaged the efforts of Boy Scouts and others in trapping the city-dwelling rabbits and transporting them back into rural areas. Thus far this method seems to correct nothing, for the rabbit population turns up again promptly in city gardens. Either the cottontails migrate rapidly onto the nearest good soil or, on the deficient, worn-out areas to which they are transported, they fail to breed properly and increase their numbers. Certainly if they have a choice between a good, well-managed farm and a poor, depleted one, they will move rapidly onto the good land which raises nutritious forage. That much we know. The ordinary cottontail however does not, as a rule, range much further than a couple of hundred feet from the spot which he has chosen as home, and this lends doubt to the theory that the transported cottontails actually migrate great distances in order to return to the suburban gardens. The evidence rather tends to indicate that rabbits transported into poor or worn-out agricultural land simply fail to live up to their long-established reputation for fecundity.

The experiments of Doctor Albrecht have shown beyond dispute the evil effects of mineral deficiencies and in particular calcium upon the breeding capacity of rabbits. He found that rabbits fed on forage grown upon calcium-deficient land failed to breed at all but that when their diet was changed to forage grown on well-balanced soil with sufficient calcium content, they bred easily and rapidly and raised large and vigorous families. The evidence worked in exactly the opposite fashion when rabbits raised on good forage grown upon good soil were changed to forage deficient in minerals and calcium. When the forage was changed these rabbits, which had been fertile and raised large and healthy families on good forage, ceased to breed at all.

Skunks, opossum and raccoons in Ohio have generally and in an increasing degree been following the pattern set by the wild rabbits. In metropolitan and suburban areas and in parks where good soil practices are observed, the populations of these animals have reached the proportion of pests. The cases of raccoons and skunks pilfering garbage cans and feeding off flower and vegetable gardens increase annually. Since in cities there are ordinances against the use of firearms for hunting purposes, the householder and gardener, plagued by the raids of rabbits, raccoons and skunks, complain to the state Conservation Commission to protect them from the wild life which prefers the easy living and good foods of the suburbs.

Not long ago in Cleveland, a skunk moved into the fruit cellar of a citizen living in the heart of the city. His presence there raised a problem for the family which had been invaded and for days the press carried suspense stories of the family's effort to evict the skunk without lingering and odoriferous results. The animal appeared to like his new habitat and spent most of his day sleeping, refusing to be disturbed. He was finally lured out of the cellar by a tasty meal of sardines and the cellar was sealed against further invasions.

Domestic animals, unfortunately and sometimes tragically, do not have the freedom to roam and choose forage which is good for them. They are fenced in and forced to eat what is grown in the fields of the farm, however poor and depleted, of which they are a part, whether it is green pasture or stored winter forage. It is undoubtedly true that the breeding troubles and the various livestock illnesses of which many farmers complain arise from the poor agriculture or the soil depletion or deficiencies of the land upon which the animals are raised. Very often a cow or a steer or a hog is forced to eat forage and even grains which are unpalatable and which they would reject by instinct if they had any choice in the matter. They merely eat it to fill their stomachs and assuage the ache of emptiness, although they may be subject to all sorts of ills arising from the deficiencies which their instinct and taste would lead them under free and natural conditions to abhor. Of course the farmer himself is the loser in the economic sense. In addition to the shame of having thin, ill-nourished and sickly animals on his farm, it hits him in his pocketbook. While the pills, capsules and injections of the veterinarian may bring temporary and superficial results, they are no real solution and serve merely to increase the losses of the farmer by augmenting his bills for pills and doctoring, while the original ills, arising from poor nutrition, continue.

To a great extent the deer in Ohio, like the raccoon and opossum, have recently taken to suburban living in increasing numbers. Ohio, save for three or four half-wild hilly counties, is not deer country like the nearby wild mountainous regions of Pennsylvania, New York State and adjoining Michigan. It is largely a big, lightly forested area of farms, orchards and nursery plantations where deer can create great damage. Not many years ago it would have been difficult to discover a deer in the whole state of Ohio, but recently deer have increased steadily in number to the point where they are not only a nuisance to farmers and fruit growers but in some areas a menace on the heavily traveled roads of this thickly populated state. Four years ago, in order to reduce the deer population, the State Commission authorized an open season on bucks, and when this failed to produce satisfactory results in reducing the deer population, females as well were included in the open season.

The odd fact is that the heaviest deer population is in Northeastern

Ohio, a thickly populated area of good farms, truck gardens and nurseries rather than in the half-wild hilly country given back to second-growth timber. Apparently, deer, like the cottontail, know where the good forage grows and are willing to undertake the menace of dogs and irate farmers in order to get really nutritious feed. And this, despite the fact that the deer is more fortunate than the smaller animals which cannot feed off the leaves and twigs of trees which, penetrating by way of their roots below the shallow depth of the worn-out and eroded soils, are able to produce forage of a higher mineral balance and quantity.

One peculiar deer problem arose in Cleveland where, well within the city limits, a small herd of deer took to living off the well-fertilized shrubs and grass of one of the largest cemeteries. The destruction they caused to the living trees and plants was bad enough, but after a time they acquired a taste for the floral offerings placed on the graves at funerals. They finally perfected the tactic of watching the funeral from a distance and waiting until the mourners drove away. Then, immediately, they fell upon and devoured the Broken Wheels, the Shattered Columns, the Gates Ajar and the blankets of roses left behind. Complaints from bereaved relatives and friends forced the Commission to take action. At first it attempted to trap the deer, but through an already established rich and abundant cemetery diet of roses, carnations and tuberoses, they merely ignored the alfalfa hay habitually used to lure deer into the traps, and eventually the herd had to be destroyed. For myself, as chairman of the State Commission I would have preferred the deer to the floral offerings, but I was regarded as unconventional by the army of bereaved and outraged and perhaps thrifty relatives and was overruled.

On our own land at Malabar Farm, as the level of the fertility was raised, and the virgin subsoil minerals made available through fertilizer, organic material and deep-rooted grasses and legumes, both the fish and game populations have increased by leaps and bounds.

Adjoining the farm on the hilly side above the valley lies an area of abandoned, worn-out farms slowly being taken over again by the forest. One could walk a whole afternoon over the area and see scarcely a dozen birds, not a rabbit and no trace even of a field mouse. But once the fence dividing that area from the restored land at Malabar is crossed, one encounters birds by the hundred, plenty of rabbits and squirrels and every sort of game including even a small herd of deer which lives with us the year round. The deer hide in the big fenced-in woodlots by day, but in early morning or in the evening they can be seen at the farm ponds, around the salt blocks and pasturing on the lush alfalfa, ladino and brome grass on which they live throughout the winter months. I doubt that they have ever jumped the line fence into the abandoned, worn-out area—more than once. There is little to eat there and what there is contains few minerals or vitamins. No chemist testing soils is a more accurate judge of the soils on the two sides of that line fence than wild animals themselves.

Any traveler by automobile is familiar with the vast amount of game killed along the highways. Some states have even erected roadsigns urging motorists to spare the game, but the real reason for the slaughter is not entirely the carelessness of the average driver and has never been posted. It is that the forage along most roads is more palatable and nutritious than that in the farmers' fields on the other side of the fence. That is so because the land along most roadsides has remained uncultivated since the first roads were laid out by a wandering Indian or a deer or by federal survey, and is virgin soil undespoiled by a poor agriculture or by the depletion of organic materials which renders the fertility of the soil unavailable to the plants and shrubs growing upon it. In the fields on the opposite side of the fence row, in the case of nine out of ten farms, the soil has been depleted and the forage is short of minerals and vitamins. The farmer's livestock is forced to eat this forage but not the wild animals which are free to wander and will reject it for the richer food on the virgin soil of the roadside.

Recently there has been a tendency throughout the country to straighten old roads which in the past, following perhaps a cowpath or a deer trail, had a tendency to wander. When the new straight road is cut across a farm field, nine times out of ten the forage which grows along the borders of the new road, cut through the depleted fields, is scanty and sickly, even to the casual glance, compared to the vegetation along the virgin borders of the old winding road. Moreover, the fertility of this virgin land bordering old roads is frequently augmented by the dust and the splashed mud coming from the bed of the road which for years had been surfaced with limestone or with mixed gravel high in mineral content of all sorts.

On the land we rent from the Muskingum Conservancy District there are two large fields separated by a county road which for two generations has been surfaced with limestone or with conglomerate glacial gravel. The prevailing winds blow at right angles to the road across the two fields carrying the dust away from the one and across the other. Although the two fields had been treated the same by their former owners and had eventually been abandoned, the fertility of the field which received regularly the blown powdered limestone and the mineral dust from the disintegrated gravel, through the years since the road was first surfaced, has always been from 20 to 30 per cent greater. It is also notable that the pheasant and rabbit population is found nearly always on the side of the road where the deposits of mineral dust had been laid down for years.

Once my partner, resenting the dust which sometimes covered the crops on the leeward side of the field, suggested inducing the county commissioners to oil or tar the roads. I promptly vetoed the proposal, observing that we were getting every year many dollars per acre worth of the most valuable minerals in highly available form without spending a penny. Sometimes visiting farmers ask whether we do not resent the clouds of dust which arise on hot dry days from our graveled farm lanes to blow across the fields. The answer is "No." The tires of the visiting automobile act as a kind of fertilizer factory, pulverizing and spreading across our fields, together with wind action, a mixture of minerals from potash and phosphorus through such valuable trace elements as manganese, cobalt and a score of others.

Where there is both good cover and good well-balanced soil of high available fertility, there you will always find the big populations of wild game, and where you find plenty of field mice, pheasants, quail and other birds, there, too, you will find the foxes, the catamounts, the owls and other predatory animals and there, too, you will find the balance in Nature operating as it should, to the benefit of the farmer, the hunter, the fisherman as well as to the wild life itself.

Ohio has led the way among all the states in emphasizing this fact and constantly conducts an educational campaign directed alike at the hunter, the fisherman and the farmer. The State Wildlife Commission pays for research and game and fish management out of the funds received from sportsmen by way of licenses and from the Federal Government which, under the Pitman Robertson Act, matches funds with the states in establishing good game habitat and conservation practices. In Ohio today most of the sportsman's money and the federal appropriations are not spent upon hatcheries and game farms which breed artificially pheasants, raccoon, fish and other forms of wild life and then distribute them over worn-out areas where neither fish nor game can propagate or even survive; the money is being spent on creating proper habitat and feeding conditions so that

[226]

fish and wild life may thrive and propagate to maximum capacity. This policy is paying off big dividends which increase with each successive year as cover for wild life increases and reforms in agriculture provide good forage and clear clean streams with a mineral content both in water and in plankton and in vegetation which encourage survival and propagation.

The old way was to spend hundreds of thousands of dollars annually to hatch and raise pheasants and then to plant them in the fields a couple of days before the season opened, which provided less real sport than the shooting of an ordinary flighty white leghorn chicken, or to hatch and raise at great cost fish which, when they saw you on the bank, followed you along in schools, in anticipation of a hand-out of liver and horsemeat. The cost of each fish raised under such a program was stupendous. In one state, a survey showed that every artificially raised and stocked trout cost $4.75 in the license money taken from sportsmen.

As expensive hatcheries have been closed one by one in Ohio, the money saved has been spent upon cleaning up streams and lakes and the establishment of dozens of headwater lakes of all sizes, both to provide sport and to create natural spawning areas for fish. About four years ago in certain streams and lakes which had been put into good habitat and propagation conditions, all fishing restrictions were lifted. Any fisherman could fish at any time of the year, taking as many fish as he liked regardless of size. As in the case of the great artificial lakes of the Tennessee Valley Authority where the same lack of all restrictions exists and the habitat conditions are excellent, the fishing has steadily improved both as to quantity of take and as to size.

This is easy to understand when one realizes that a single female bass or trout produces up to two hundred thousand young fry. Crappies, sunfish and bluegill or bream are even more prolific. In some lakes and ponds the fishing is poor because there are too many fish and they are all stunted and undersized because there is not enough food for them. A fish does not grow according to age but according to how much it gets to eat. A five-year-old bass may be seven or eight inches long or a three-year-old bass twenty to twenty-four inches according to the abundance of its diet. Considering the fecundity of any female fish, there should be and in fact there is no need to stock artificially a stream, a lake or a pond where conditions of food and habitat are good.

A few years ago at Pleasant Hill Lake I happened upon a State Commission fish truck dumping young hatchery bass five or six inches long into the clear water of this lake which borders our farm. He did not know that I was a member of the Commission, and in the course of our conversation he observed, "This is the damnedest foolishness I ever saw. It cost a small fortune to raise these fish and a pair of bass will produce for free about fifty times as many fish as I'm dumping today." I hastened to add that I agreed with him heartily.

Ohio is a thickly populated state, largely agricultural with great industrial cities scattered evenly over its area—Cleveland, Youngstown, Columbus, Cincinnati, Dayton, Toledo, Akron, to name a few, with dozens of other cities with populations ranging from fifty to a hundred thousand inhabitants. But Ohio has some of the finest fishing in the world, much finer in fact than that in many states which spend millions each year advertising their fishing to attract the tourist and the sportsman. And each year as Ohio's agriculture and forestry practices improve, as new lakes are constructed and streams are freed from pollution, the fishing grows better.

A good deal of my life has been spent in Europe in areas more thickly populated than Ohio where there was *natural* game and abundant fishing in such quantities as I have rarely seen in the United States. Both fish and game existed even up to the suburbs of large cities and in smaller towns one could have excellent trout fishing in the streams which paralleled the main streets in the very center of town. In these areas in France, in Austria and in Germany, this abundance existed because conditions of habitat and the regional agriculture were excellent. There was good cover and there were clean streams and there was good, well-managed soil which produced quantities of good forage. In Ohio we are beginning to approach such conditions despite the size of our cities and the great numbers of people in need of recreation and sport. Many an Ohioan can find better sport today within the borders of his own thickly populated state than he will find by journeying hundreds of miles to states which make much more of a hullabaloo about the hunting and fishing that is to be had within their borders—and make much less effort to create the conditions which provide top-quality sport.

In all of this, good soil and clean water, and the minerals and vitamins that go with them, are of prime importance. Next comes the farmer who is the second most important factor, for in his hands very largely lies the control of such conditions. Cities and industries may pollute streams and lakes and do, but by far the greatest cause of the depletion of fish populations and of poor fishing is the pollution

[228]

which comes from siltation and the erosion of soil from badly managed farms in the hands of ignorant or stupid farmers.

Game fish will not live in muddy, flooding streams, and siltation in the form of mud eventually prevents them even from propagating. First the game fish go, to be replaced by the sluggish carp and catfish, and then, as in some parts of the eroded Deep South, there are no longer even carp and catfish but only turtles and snakes.

The conditions work backward as well, up the streams onto the land itself. The badly managed farm not only loses its topsoil and the availability of its mineral fertility, but the farmer's income and economic and sociological status go with it. And each year there is less and less wild life as game animals desert the wretched forage grown on poor ill-managed land for the richer stuff, which they recognize by taste and instinct, that grows on good, well-managed and productive soil. All of these things—good hunting and good fishing, good crops and abundant propagation, whether of wild life or domestic animals, come from the same place as the farmer's prosperity and independence and self-respect—out of the earth.

CHAPTER XII

Why Didn't Anyone Tell Us?

When shall we have an entomological laboratory for the study not of the dead insect, steeped in alcohol, but of the living insect: a laboratory having for its object the instinct, the habits, the manner of living, the propagation of that little world, with which agriculture and philosophy have most seriously to reckon?

—EDWIN WAY TEALE,
The Insect World of J. Henri Fabre.

XII. *Why Didn't Anyone Tell Us?*

THERE are many hundreds of thousand of good farmers in the United States and many more hundreds of thousands of bad or indifferent farmers. It would be difficult to imagine any one factor more important and more beneficial to the whole of our economy and even of our civilization than a population of farmers who were all good and intelligent husbandmen, independent in character and prosperous and secure in life through their own efforts. It is immensely encouraging that the breed of farmer is constantly improving, away from the indifferent farmer of the frontier and the vaudeville "hick" who inherited the declining or "run-down" land wrecked by his predecessors during the end of the nineteenth and the beginning of the twentieth centuries. It is immensely encouraging to discover everywhere young men and women growing up in the warm new tradition of the 4-H Clubs, the Future Farmers of America and the Vocational Agriculture classes.

One of the most stimulating experiences in a lifetime fairly rich in stimuli has been the contacts we have had at Malabar with countless G.I. training classes. These are the "on-the-job" training classes set up under the Veterans Administration to give the G.I. both instruction and practice in the profession of farming. I use the word "profession" advisedly, for the atmosphere surrounding these groups of G.I.'s, sometimes several to a single county, is one of dignity and of stimulation on a higher level than that known in the past in American agriculture save among the limited number of naturally intelligent and persevering husbandmen. For the most part, the veterans are in their late twenties and early thirties, with a seriousness of character and purpose which is, at times, very moving.

The level of teaching, from men selected by the G.I.'s themselves and approved by the Veterans Administration, is singularly high, not necessarily in the academic sense because there is no specification of advanced college degrees, but because most of the instructors are close to the earth and to animals, love their jobs, are proud of their boys and are more often than not successful, practicing farmers themselves. This is immensely important in the teaching of agriculture and immensely important in relation to the things I have to say further on in this chapter. These instructors have their hands *in the earth*.

One does not learn about soils and animals exclusively in a classroom and one does not maintain the vitality and sparkle of the great teacher by passing years in an armchair and working out formulas without ever putting foot on God's good earth.

The farmer eternally believes what he sees and learns from the reality of the world around him and from his own observation. Merely to teach a farmer how to observe and what to look for is, in itself, the greatest possible achievement and one which is far too greatly neglected in our present system of agricultural education. I have found few things more humanly encouraging and satisfactory than the light which comes into the faces of some of the farmer instructors of the G.I. training classes as their "boys" see and hear new things about farming and animal husbandry. It is a light I have seen too rarely in the faces of the agricultural college instructor or even in the visages of many county agents.

There is in the faces of these farmer-instructors none of the boredom or condescension that I have encountered sometimes in other fields of agricultural education. There is none of the smugness that too often accompanies a university degree acquired through "cramming" with a minimum of inspiration. In the face of the instructor there is respect for the intelligence of his "boys" and their eagerness to be good farmers and to get ahead in the world. He is not merely holding down a job, for his payment is ridiculously small in relation to the amount of time and energy he expends. He is really trying to do a job and as often as not has the spirit of a crusader. He understands, sometimes without knowing, what it means to "spark" a young fellow so that he will carry on for the rest of his life with a lively interest in all the phases of soil and agriculture long after his "on-the-job" training is finished.

The spark, I think, is the thing that is most overlooked in our agricultural training. The lack of it, and the lack of capacity for producing it, marks the greatest failure in our agricultural education, and this failure I think arises partly from the fact that the average teacher in agriculture does not sufficiently respect the profession to which, somehow or other, he has attached himself. There is nothing sadder or more discouraging than the agricultural college professor or instructor, sitting in his office with his feet on his desk, killing time between classes of young people who bore him because he has never found his profession exciting or has never bothered to learn anything new since he received his degree.

My maternal grandfather, of whom I have written many times, and who died at eighty-six one of the best-educated people I know although he never went to school after the age of eighteen when he ran away to the Gold Coast, once said to me, "If a person really wants an education, there is nothing on earth that will keep him from getting it. If he doesn't want it, nothing on earth can impose it upon him. But the fact is that the great majority of the human race lies in between. They are average and they don't know whether they want an education or not and all they need is a sparking. It's like striking a spark from a flint to tinder. Start the fire and it will go on and on opening up constantly new excitements, new curiosities, new stimuli for the rest of an existence."

What he said applies more to the potential farmer than perhaps to the men and women of any other profession. A little "sparking" is what is needed rather than a cramming with many facts which are frequently of little use to him and many times are obsolete or proven unsound by the time they get to him.

It need not always be young people who need the sparking. I have seen middle-aged and even elderly farmers who had been practicing an indifferent kind of farming listlessly and without interest in the "grandpappy" pattern until suddenly a county agent or a neighbor or a son or daughter made apparent to them that they were practicing the most exciting and satisfactory profession in the world without ever knowing it.

I have never forgotten the look in the eye of a sixty-five-year-old hill farmer in the poorest county in Tennessee, who could neither read nor write, as he showed us the rich grass and legumes that had replaced the broom sedge and poverty grass on his steep fields, and made us, in his excitement, scramble down and climb back up a half-mile hill in order to see the fat white-face cattle which were lost somewhere in the brush of his bottom pasture. He had been sparked at the age of sixty when, through some quirk (perhaps because he thought he couldn't be any worse off), he had volunteered to work with the Agricultural Agents of the T.V.A. in setting up a pilot farm. There was no doubt that the last remaining years of his life would be richer and more exciting than all the rest of his life put together. His bright blue eyes had a light in them that was never there even when he was a boy of eighteen. Some T.V.A. agricultural teacher had set a spark to the tinder.

Although no exact figures are available, it is probable that the nation, including federal and state colleges and agencies of all sorts, spends in the neighborhood of $3,000,000,000 a year in aid and education for farmers. No other group in our society benefits in anything like the same degree and the reader might well ask, "If so much money is spent every year, why has so little progress been made in improving our agriculture?" The answers are, I think, almost endless.

There are in the nation forty-eight agricultural colleges more or less controlled and financed by the states themselves and there is the vast Department of Agriculture, until the recent war our most expensive department of government, and the Extension Service, operating more or less jointly between the Department of Agriculture and the various state colleges. And there are many other agencies of education such as the Soil Conservation Service and the rather political Triple A, now become the even more political and bureaucratic Producing and Marketing Association, and there are Farmers' Institutes held locally in winter and short courses for farmers at many state colleges during the winter months. And there are thousands of truck-loads of pamphlets upon every subject from agronomy to teaching the farmer's wife how to make a bed. And there are numerous agricultural foundations privately endowed and films made by industrial companies and private foundations and distributed free of charge in an enlightened self-interest.

Out of all this, it is astonishing what a small trickle of information, much less of inspiration, finally comes through to the average farmer. The tragedy is that the poorer the area, the poorer the farmer, both in knowledge and economic status, the less the information that reaches him. This is particularly true of the 60 per cent of our farmers who produce very little more than they consume and acutely true of the poorest regions of the South and the Southwest where actually there exists suspicion and hostility of most agents associated with government of any kind. In such areas the local church is a potent factor, and if every local preacher, most of them of the "shoutin'" evangelistic variety, could be enlisted as a force for good agriculture, good diet and good soils, the advances made both in health and in economic terms in such areas would be tremendous.[1]

[1] In recent years there has actually come into existence an organization known as The Friends of the Soil which works with the local preachers in bringing new agricultural ideas to the people of backward areas. It has even developed a ritual dealing with the soil, its meaning and its potentialities which is used in church services. The ritual, of course, is sprinkled with the innumerable and beautiful quotations relating to husbandry and the soil which abound in both the Old and

The good, intelligent farmer, the "sparked" farmer will somehow find the information he wants, but often enough he has to dig for it.

Part of the answer lies probably in the old and classic statement that "teachers are born and not made" and that there are all too few real teachers in the field of agricultural education, or any other field for that matter, with the capacity for "sparking" the students working under them.

At top, of course, the principal cause of the failure is bureaucracy and red tape and politics, all of which demonstrate their evil operations in a hundred ways. Sometimes the dean of an agricultural college is appointed politically by a board of college directors who in turn were appointed politically and none of whom have much knowledge concerning agriculture or agricultural economics. I have known one dean, so harassed and preoccupied by politics and administration (campus politics as well as party politics) that he did not even know what was being grown on his own farm. It is obvious, I think, that he had small contact with farmers and small time to devote to directing the kind of education which would be profitable to both individual farmer and nation.

And there is the kind of "gentleman" dean who is more concerned with speaking at dinners of industrialists and bankers than with the farmer and the soil from which he becomes eventually so remote that his contribution is virtually nullified. And the kind of dean who does not want to extend his various departments or enter new fields of research because it will cost money which he has to beg from a frequently agriculturally illiterate board of college trustees or directors, or because, in the interdepartmental jealousies and politics, any action in a progressive sense is likely to raise too many rows. I know one dean who, when asked what the college was doing regarding the study of trace elements, responded, "Well, we haven't done anything yet but we're being forced into it." Forced into it by whom and by what? By the good farmer, the privately financed agricultural foundation, the medical profession and the agricultural magazines which do a heroic job in helping the farmer, and even by the fertilizer companies. "Forced into it" when an agricultural college should have been leading the way to prove either the virtues or the fallacy of the effects of

the New Testaments. The Friends of the Soil is not the same organization as The Friends of the Land, a society which has done much invaluable work in propagandizing the importance of our natural resources and their conservation throughout the country during the past ten years. The two organizations, however, work closely together.

trace elements. Then he added lamely, "But if I authorize any special research, it will make an awful row because some of my professors don't believe in trace elements." A fine basis for advancement of science or education!

In the same field I recently heard a professor say, when asked concerning a trace-element product marketed by one of our large industrial corporations, "Oh there's nothing to that! No use spending money on that stuff," and then almost immediately catching himself, he added, "Oh, maybe there is after all. That company is paying for three fellowships here at the college!" When asked how he knew whether the stuff was any good or not or whether he had ever tested it, he answered, "No, of course not. We haven't time for things like that."

Meanwhile, Johns Hopkins University, a non-agricultural institution concerned mostly with medicine and one of the first-ranking universities in the world, has allotted $500,000 to the study and coordination of already existing information, in the same field, and Cornell University, with perhaps the greatest agricultural school in the world, last year set up a whole new School of Nutrition with one department dealing with the relation of minerals to the health of plants, animals and people.

Missouri University and State Agricultural College are also making valuable contributions to the field of minerals and nutrition and crop yields, largely under the leadership of Dr. William Albrecht, one of the great authorities on the subject.

Recently another college, when requested by the state Wildlife Commission to compile information regarding the relation of poor and depleted soils to wild-life population, requested ten years and $40,000 to set up a research project although the information already existed and much of the work had already been done by other colleges and the Wildlife Service. The problem was merely one of coordination of already known material with no need whatever for going over the whole field again in an expensive process of the wholesale duplication so widespread under our disorganized system of agricultural education.

To date most of the information in the whole field of minerals and nutrition has come from industrial foundations, medical colleges and good farmers endowed with powers of imagination and observation. The medical profession as a whole has displayed a steadily increasing interest in the so-called trace elements, especially in relation to the endocrine system, as medicine has moved increasingly toward the

[238]

philosophy of *preventing* disease and physical breakdowns rather than sitting by until they have occurred and then trying to patch up the patient.

Successful Farming, one of our best agricultural magazines, has recently published an illustrated booklet called *Getting a Better Living from Your Soil* which is in itself an agricultural education with all of the nonsense removed. It is streamlined and provides within its hundred pages the fundamentals of a good and modern agriculture. Virtually every article is written by individual specialists and members of the faculties of agricultural colleges and the level is uniformly and remarkably high in intelligence, in scientific value and in readability. Indeed, any layman could read the whole booklet with great interest. Yet the remarkable fact is that somewhere along the line, owing to deficiencies and handicaps of bureaucratic education, very little of the same information ever finds its way through the intricacies of a system down to the level of the farmer. It is as if, in order for these admirable men to reach the farmer, they are forced to escape from the system and write for magazines or to talk to the farmer personally.

The same high level of sound education holds for all of our best agricultural publications in which very often the farmer can find much more information of a fundamental and valuable quality than he is able to get from the state colleges or through the Extension Service. Perhaps the great difference is that in the magazines, the editors respect the intelligence of the good farmer and explain to him actually what happens in the earth and even in his livestock, presenting it in a way which makes the material not only interesting but fascinating and stimulating. Of course it is not filtered through endless offices and bulletins and the censorship and prejudice of many small men, but goes direct from the scientist to the farmer through the medium of the magazine. One often gets the impression from reading our farm magazines that the editors and the contributors, who frequently include top-level dirt farmers, are far out in advance of many of our state colleges, yet, curiously enough, the best articles are sometimes written by scientists from the very staffs of the colleges which seem most backward.

What is true of the *Successful Farming* booklet is true also of the excellent Yearbooks put out by the Department of Agriculture each year, where the individual professor writing direct to the farmer does the kind of job which is needed. It would appear in the case of the colleges that too much of the writing, too much of the instruction, is left in the hands of bored or limited bureaucrats or instructors and little men all the way down the line or in some cases these things have suffered from the limited abilities and prejudices of deans and directors. The same excellent qualities exhibited in the books mentioned above hold true of another book which no real farmer should be without. It is *Hunger Signs in Crops*, published by the National Fertilizer Association of Washington, D.C. Again the professors write brilliantly on their own when writing directly to the farmer. It would appear that the college campus enveloped them in a fog of mediocrity, dissension and apathy.[2]

And what happens to the individual teacher or research expert with the brilliant mind and the capacity to "spark"? More likely than not one of two things happens to him. In a welter of campus politics and jealousies, the armies of mediocrity and laziness gang up on him, or his brilliance and his achievements are recognized by some industry or privately endowed foundation or perhaps by a better agricultural school and he is hired away at much more money than that provided by his existing inadequate salary. In too many of our agricultural colleges the faculty, under such combined pressure, eventually becomes a standard of mediocrity but of little else, staffed by men who are able to remain there only through an untroublesome conformity and mediocrity, teaching on and on year after year, frequently out of textbooks and theories long obsolescent or enveloped in a haze of academic monotony and boredom.

Certainly the costs and the inconvenience of prescribing new textbooks is one factor in the general mediocrity of agricultural education. There are books being used in some of our agricultural colleges which not only became obsolete long ago but are actually teaching methods and ideas long since proved not only false but actually destructive to a good agriculture. It has been too much trouble or no one has cared enough to eliminate such books or bring them up to date and in line with the fabulous discoveries made in the whole field of agriculture within the past few years. And, of course, there are still a good many professors and instructors, as out of date as the textbooks, who have never learned a new fact nor asked an imaginative or scientific question of themselves or of the students they teach since the day they achieved their college degrees.

And occasionally the dean, who is frequently subject to the

[2] The University of Oklahoma Press, under the direction of Savoie Lottinville, deserves special praise for the fine list of stimulating and valuable books on all phases of agriculture which it brings out from time to time.

politically appointed, agriculturally illiterate board of directors, cannot get the appropriations he wants to run a real college of agriculture. Such a college, and even education itself, frequently becomes wholly unimportant to the directors in comparison with the new $3,000,000 stadium or the bargaining with the Peanut Bowl or the Soybean Bowl or some other commercially managed athletic field over how much they will get for the appearance of the Siwash team in a contest with the Freshwater team.

And our federal Department of Agriculture, even after a fairly thorough housecleaning by Milton Eisenhower, is still an incredible monstrosity with one division and bureau overlapping another and creating in the process all kinds of duplications, feuds and jealousies. Even the Hoover Report understated a confusion which ranges from the divisions of research through the endless economic agencies set up all too often merely to buy the farmer's vote. Feuds, jealousies and name-calling arise out of the muddled and conflicting authorities which not only nullify the purposes for which they were set up but frequently create actual harm for the whole of our agriculture and economy.

Certainly these difficulties are of little help to the farmer looking for information. More often than not the spectacle of one bureau or one individual feuding with another merely confuses him and turns him cynical regarding the whole department. The very pamphlets sent out to him are frequently unreadable because of their dullness and lack of clarity and frequently enough they contain information actually contradicted by the next pamphlet that comes along. The purpose for which the Department was set up—that of research and instruction to benefit the profession of agriculture—has at times become lost altogether in propaganda for or against this or that political issue, for or against this or that division or service or bureau. Indeed, no Department of our government (and only the Defense Department is more expensive to the taxpayer) is so riddled with the evils that increase as bureaucracy mounts and mounts. Moreover, the post of Secretary of Agriculture, set up originally to direct research and agricultural science, has become increasingly merely an office for disseminating party political propaganda to buy the vote of the farmer.

The Extension Service, operating jointly with the state agricultural colleges to bring information and benefit to the farmer, again suffers from the evils of red tape and of bureaucracy. The county agent, often ill paid and devoted and able, working in the individual county, is forced to do acres of paper work, filling in reports and questionnaires which are forwarded eventually to those vast archives in Washington so costly to the taxpayer and so handicapping to the farmer, where they repose without even having been looked at to gather dust as more and more buildings are erected to house them. Some perhaps are never used for any purpose whatever.

This red tape, demanded by the Federal Government and sometimes by the state colleges, is one of the greatest barriers between the farmer and the information he desires and needs. It occupies many hours a week of the county agent's time which otherwise could be spent in the fields on the earth with the farmers of a county. This, oddly enough, was the original purpose for which the county agent came into existence but it is a purpose which sometimes seems to be utterly forgotten by the bureaucrats in their dusty cells. The county agent was designed to be the most direct link between sources of agricultural information and the farmer but, as such, the county agent is handicapped in many ways of which bureaucracy and red tape is but one. He is subject to the censorship of the state college and sometimes to the authority of a fire-eating, unscientific, political dean or Director of Extension who sets up his own prejudices rather than scientific fact or research as the measure of information which should reach the farmer.

Many a county agent, in his practical and even scientific approach to agriculture, is far more advanced and accurate than some of his superiors. And frequently he becomes the victim of feuds and jealousies occurring on higher levels. Not the least example is the kind of feud which exists in certain states between the Extension Service and the Federal Soil Conservation Service, a feud promoted by certain state Farm Bureaus, agencies which have no legal standing whatever in government but which in certain states have actually used unscrupulous pressures to turn the Extension Service, paid for by the taxes of all of us, into a recruiting service for increasing their memberships. The feud, needless to say, thrives most in the states where, generally speaking, there is a high proportion of the absentee landlordism which has been for long so costly not only to the natural resources of the nation but to the individual taxpayer as well.

All of these difficulties, vices and failures are perhaps not more than the results of the human failings of mankind, aggravated and magnified when bureaucratic government increases, but the farmer and the taxpayer are the victims of them all—the farmer in terms of the partial failure of a system to help him and the taxpayer in the

[242]

huge sums he must pay for the working of a machinery which is faulty and confused and duplicating in its operation. Many excellent results come out of it but, as often as not, the results come by accident or through the outstanding qualities of the individual, from county agent to director or college dean, who manages somehow to survive his infinite restrictions and handicaps and break through to the farmer himself and to that agriculture upon which so much of our economic salvation, health and vigor as a nation depend. Somehow, in all the red tape, in all the jealousies and feuds, duplications and human weaknesses, the capacity for "sparking" the farmer becomes dimmed and often lost. It survives only with the officials who have courage and intelligence and perhaps the diplomacy to carry out their purpose despite the weaknesses both of system and of individuals, but under a bureaucracy which tends to reduce all human talent to utter mediocrity, they do so at a decided risk.

They do fortunately exist—these individuals who manage somehow to carry through with the purpose to which they are dedicated. They exist among the deans, the professors, the county agents, but almost without exception they are able to accomplish their high purposes not because of but in spite of the politics and bureaucracy and feuds and prejudices and red tape of a system. They are the heroic ones and they do carry the spark. One of them is worth fifty of the bureaucratic or academic conformists both in character and intelligence and in dollars and cents to the taxpayer and the farmer.

In the services of the Department of Agriculture and in the Extension Service and the state colleges there exist men of the highest ability and devotion, frequently serving for salaries representing a fraction of what they are worth in terms of intelligence, character and training. To them the farmers and the nation owe a very great debt. Some have attained recognition in the areas in which they work and some even beyond the limits of those areas. In one sense the vast Extension Service is as good as the local overworked county agent on whose shoulders fall an increasing number of duties and obligations, many of them conceived and imposed by some dusty statistician in an office hundreds or even thousands of miles away. I have known county agents who have made over the whole economy of their counties through their own personal efforts, and they have done so frequently at the risk of losing their ill-paid jobs because they have paid more attention to soil and agriculture and the farmer than to the endless silly forms and reports they are expected to fill out and turn in.

[243]

The job to be done lies *in* the soil and *in* agriculture and not merely in reporting this or that tidbit of useless information to some remote and sterile statistician.

Certainly one of the great errors of our agricultural higher education at the present time is the exaggerated emphasis placed by many state colleges on producing only technicians and specialists. The error no doubt arose out of the need a generation or more ago of more experts and more teachers to go into the field and work with the farmer. That need is long past, although the need for *top-quality*, inspired teachers is still very great, as indeed it probably always will be. But at the present time there is a glut of technicians, of specialists in every field of agriculture, so that many a young man or woman emerges from an agricultural college with a diploma and no place to go. Moreover, all too often the young man or woman came from a farm and would make perhaps his greatest contribution by going back to his own area and doing a first-rate job, but he has in a sense been spoiled as a good farmer by a highly technical education as an "expert" and technician and has acquired a curious kind of snobbery regarding the soil and fundamental agriculture itself. And sometimes the pressures and overemphasis on the particular field in which he has chosen to specialize have thrown him out of balance altogether so that he falls into the error of so many specialists and believes that *all* the answers to *all* the complex problems of agriculture can be found in his own narrow field.

In no field of our education, unless until quite lately in the field of engineering, has overspecialization been so rampant as in agriculture. One has only to look about in the field of agriculture to find any number of normally intelligent young men and women who might have made excellent farmers and prosperous productive citizens of the nation but have been in a sense deformed and thrown off the track into an overcrowded field of activity where it is becoming increasingly difficult to find a good job.

At one of our greatest state agricultural colleges only 4 per cent of the 1948 graduating class expressed the intention of going back to the farms from which they came. The others were all technicians and specialists setting out into an already overcrowded field.

This same system of attempting to force young people to become technicians and specialists rather than good farmers shows up throughout the study plans of most state colleges. On the assumption that *every* student wishes to become a specialist, the tendency is to impose upon each and every one the necessity of choosing a major subject and then directing his education in that direction. As a result, many young people who want to learn the fundamentals of good agriculture as quickly and as efficiently as possible are forced to spend countless hours on higher mathematics, on advanced chemistry, even on foreign languages which will never be of the faintest use to them, and to spend much time and money taking a four-year course of which 50 per cent is unnecessary and often without value. I have, indeed, known many young people of great intelligence and much valuable experience as farmers who after a year or two became utterly discouraged and left college because of the long hours and schedules designed for "specialists."

Here again the question of "sparking" is of the utmost importance. What is crammed into a young head out of books is of no importance if it is unaccompanied by inspiration and it will be more quickly and easily forgotten than learned. The student trained in fundamentals and "sparked" will find out during the rest of his life all he needs to know regarding his profession and he will continue to find out until the day he dies. The "short courses" offered by some state colleges during the winter months have been apparently the only concession made away from this tendency to force down the throat of every student a mass of stuff which is both meaningless and valueless, however much it may please the individual professor teaching it. Many a young person cannot afford four years of expensive agricultural education in terms of money, but many more, impatient to get started in life, cannot afford the fancy work in terms of time and discouragement.

In some cases the Extension Service has been guilty of pushing contests and projects into the field of the unsound and even of the absurd, especially in terms of sound regional agriculture and livestock raising. Certainly one instance is that of feeding animals for prize 4-H awards far beyond the limits of economic soundness. One outstanding example, pertinent because of the exaggeration, occurs occasionally in the fat stock contests conducted in some of the Gulf Coast counties where young people bring in calves from other areas and then feed them on a regional basis suitable to Iowa but wholly unsuited to the cattle industry in a warm, humid, insect-ridden country where corn is not a profitable crop. Quite obviously the prize should go to the young person who raised a Brahma or a cross-bred Brahma in the terms of the community in which he is living, rather than bringing in a breed unsuited to the climate and the forage of the area and then raising him as if he

[245]

were being raised on a feedlot in the rich corn-belt area of the Middle West where corn may be raised efficiently and cheaply. The winner sells his prize steer for several thousand dollars of businessmen's money and the rest of the contestants are left holding the bag unless some sympathetic oil man comes along and pays for the expensive feed which has gone into the cattle they have been raising and feeding. Fortunately for the contestants this is usually the case, but the whole procedure has little to do with efficient and profitable cattle raising in an area which essentially and continually is a breeding rather than a feeding area for cattle.

A similar error regarding the relationship between animal husbandry, pure-bred cattle and good feed and forage exists in many of our agricultural colleges where in some cases 75 per cent or more of the total money allotted to these things is spent in teaching about blood lines, registry and judging animals as against 25 per cent and less on nutrition, on good agriculture and the raising of good pasture and forage which are the fundamentals as compared with which the rest is merely dependent and accessory. The same error of course extends into the field of pure-bred cattle activities as practiced by many city farmers and some practical registered cattle breeders who should know better. In all too many cases, a whole staff is maintained to keep track of pedigrees and tests of $25,000 bulls and $5,000 cows while the animals themselves are frequently on third-grade pasture and forage or being fed unbalanced, forced grain diets. To be sure, the bad results are inevitable in shy breeding, in a variety of diseases and, in the case of many valuable dairy cows, in "burning out" the animals in a few years. Veterinarians are brought in and give the animals injections and capsules in order to cure shy breeding and other ills which should never have occurred in the first place if the owners and often the college-educated managers and specialists concerned had had the opportunity of a proper and balanced education. In the same school of thought are those "record-breaking" dairies which feed $2 a day worth of feed to get 5 cents more milk, in total neglect of that economic line which means profits for the farmer and health and longevity for valuable cattle.

Indeed, the whole emphasis on line and in-breeding has pushed in many breeds to the point of absurdity, both physiological and economic, and in many cases to the deterioration of qualities of ruggedness and high production on cheap good forage which are the essence of profit for the farmer and commercial cattle breeder.

This has been increasingly so in the range of show animals in some of the beef breeds where the process has been shrewdly and brilliantly described by an experienced old rancher friend in Texas. He says,

First you give the bull hormone injections and then aid him to breed the cow. Nine months later, you summon a midwife to deliver the calf. Once on its feet it spends half an hour trying to find the udder of the cow and when it is at last located, the calf gets a half cup full of milk and you end up by bringing in a Holstein to feed and raise the calf. A year later you help the calf to his feet, put him on roller skates and crimp his coat, shine him with oil and wheel him into the show ring where he wins first in the yearling class, and as soon as he is old enough to breed, the process begins all over again. But what in Hell has this to do with the kind of cattle which can scrounge a living and bring up a profitable calf on the range? Maybe it may please some millionaire or some judge who has never made a living raising cattle but there ain't much sense of any kind otherwise.

In this overemphasis on breeding, on pedigrees and on records I have known many a small general or dairy farmer to lose his shirt. Indeed, in some states one might gain the impression that there was something disgraceful about good, healthy, productive grade cattle operating at a good profit to the farmer. All the pressure is toward every little farmer having a registered and accredited herd, often enough regardless of whether they are feeding on weeds or on good forage. As a corollary of this, the farmer is encouraged to do heavy-grain feeding and to spend his potential profits on every kind of protein and mineral concentrate and supplement. He must go to the expense of a professional official tester and record keeper and spend hours over pedigree papers, when all the time he could be making a lot more money on high production per acre of good high-protein, high-mineral-content forage while operating a good, selected grade herd of animals.

The sad experience of many a dairy farmer lured into a registered cattle program is again best told in the words of an old-timer friend from Iowa who has been through the whole racket. He puts it thus:

To really get anything out of the registered racket you have to be in the top level. My father's herd used to average at sales $1500 and better an animal. When he got old and the operation fell into the hands of my brother it went to pieces because he didn't stay in the top level. When my father was alive he spent hundreds of thousands of dollars in showing his herd, transporting it from one fair to another. His only rival was

Mr. X who had just as good a herd and they worked the racket together. One year my father would import from England with great fanfare a $20,000 bull, and breed him to his best cow. When the first bull calf was born, Mr. X would buy it from my father with great fanfare for $2000 to $3000 and give him a note in exchange. Then the next year Mr. X would bring in with great fanfare a $20,000 bull and breed him to his best cow and as soon as the first bull calf was born my father would buy it from Mr. X and give him a note for $2000 to $3000 in exchange. Then both of them tore up the notes and suckers who came to the sale bid on that level of prices for the calves and paid for the two $20,000 bulls and maybe my father and Mr. X showed a profit.

All too often the average farmer with a pretty good and profitable dairy herd is urged to go in for registered cattle, and he goes out and spends $5000 for a bull and maybe $500 or $600 apiece for heifers. The prospect, according to his advisors, is now brilliant although he has added to the expense of his production through testing, special fancy feeds and in a great many other ways, and the chances are that he has less vigor and even perhaps less milk-producing capacity in some cases than in his old grade herd. Then when his calf crop arrives, the magic seems to vanish. He has purchased heifers for $500 or $600 apiece, but when he attempts to sell such heifers back to the big registered herd buyers, the most he can get for his heifers is little more than he could get for one of his grade heifer calves.

There is no question whatever of the great value of the registered cattle of this country nor of the good and scientific work that has gone into their breeding. The question again is one of getting off on the tangent of overspecialization and of breeding lines so close in order to produce a "pretty" animal which pleases the current fashion of the show-ring judges that vigor, profitable feeding capacity and many other qualities become lost to the detriment of both dairy and beef industries. And the man who owns and operates profitably herds of registered cattle must have a certain temperament, a certain pride and even perhaps an obsession for the registered animals. It is not a business for the average farmer, and to urge him into it on the basis that grade cattle are a disgrace and that he will make more money from a registered herd is doing injury both to the individual farmer and to agriculture and the livestock industry in general, especially when the fundamental factors of a good agriculture and good and cheap forage are overlooked in the rustling of the pedigree papers under the lamp at night.

[248]

Occasionally I go on the so-called "feeding tours" through the corn-feeding lots of the Middle Western states. It is a tour usually under the direction of state college officials to observe beef feeding practices and discover which ones are the most profitable. Often enough, perhaps as much as 50 per cent of the time, the most money is actually made not by the farmer feeding pure-bred beef cattle but by the farmer feeding Holstein or Brown Swiss steers or steers of those breeds crossed with some beef breed. The reason is simple enough. The big dairy breeds and the cross-breds cost less to buy in the beginning. Because of their big frames and rapid growth they put on, in bone and meat together, a much greater poundage of meat per pound of feed or per acre of pasture than many a pure-bred beef steer bred to produce top-quality cuts of meat. When marketing time comes the dairy or cross-bred steer feeder takes less a pound for his animal at the stockyards, sometimes a lot less, but in the greater weight of his steers he gets a bigger price and a great many times bigger profit. One Kansas feeder told me, in two feedlots of steers—one cross-bred Holstein and White-face and one pure-bred beef cattle—his cross-bred steers brought him at the peak of the market in 1948 $80 a head more than the pure-breds.

This is, of course, talking economics and not cattle breeding, and quantity not quality, but in recent years the quality beef has attained a price so high that the market has become restricted and the spread between the price of top-quality beef and that of second- or even third-grade beef has narrowed to a point where many a feeder no longer aims at the top-quality market but at the level which will make him the most money, with animals which make the most rapid efficient growth in a market which shows the greatest demand. Under such conditions we again approach the law of diminishing returns and the fact that when any agricultural product or commodity costs too much to produce or is priced too high for other reasons, the market begins to contract, and a surplus is created or prices fall to a point where demand again becomes active. It is a rule and a law on which the colleges might well place greater emphasis to the benefit of countless farmers who could make much higher profits than they are making, not from higher prices, but from more efficient and productive operations at home on their own land.

The country banker, like the country preacher, has a great role to play in the improvement of our agriculture and consequently in the improvement of the whole economy of the nation; and increasingly the country banker is carrying out his responsibility. Not always has he done so out of mere disinterested virtue but because, tardily, the banker, from the small state bank to the great banks of New York City, has come to understand how much a sound agriculture, upon which all our economy is largely based, means to the banks and to the nation. He has seen country banks die like flies for two generations and he knows out of bitter experience that when an agricultural area dies, the local bank dies with it. It no longer has either depositors or borrowers.

Today countless country banks have engaged agricultural specialists not only as advisors in making loans but as actual teachers and consultants. If a farmer is persistently a bad farmer, he will deposit no money and the bank cannot afford to loan him money; but the bank itself can aid and counsel him and even, by making restrictions on a loan, force him, sometimes against his own stupidity and stubbornness, to farm well and prosperously with proper regard for his land and his livestock.

In Tennessee, C. W. Bailey has made over the economy of seven counties from poverty to prosperity simply by his advice and his power as a country banker. In Arkansas, W. W. Campbell has made the same kind of contribution to the nation. Even banks as large as the potent Wachovia Bank in North Carolina, under the leadership of Robert Hanes, are carrying out statewide programs of farm education and aid, and the district banks of the Federal Reserve System have all instituted notable agricultural programs of which that of Chester C. Davis, President of the St. Louis Federal Reserve Bank, established the precedent. With all of these forces at work, it is only logical to ask, "Why have we so many bad farmers?"

I think that the farmer, too, must shoulder a large amount of the blame, sometimes because he is lazy, sometimes because he is overworked, sometimes because he is ignorant and chooses to remain so, sometimes because he is trying to make a wretched living on land which is not agricultural land at all but semi-desert on which no man, however clever or informed, could make ends meet, sometimes because he belongs in that abominable tenant-absentee landlord combination in which both landlord and tenant are squeezing from the dying soil its last traces of fertility. And sometimes the farmer is merely too old to take any interest in improving his land or his income. He is merely waiting to die.

Fortunately there are great changes in progress, arising principally from the younger generation which has been given a different point of view regarding agriculture. Largely speaking, their point of view represents the New Agriculture in which the farmer is part businessman, part specialist and part scientist rather than the old, wasteful, ignorant, frontier agriculture or even an agriculture in the four-year rotation general farm pattern. This change will come about eventually through the force of economics alone since the older patterns are profitable neither for the individual nor for the nation. Moreover, they imply two other defeating factors—an overburdening investment in different kinds of machinery and long hours of drudgery.

There is a good deal of sentimentality awash throughout the nation with regard to "farming as a way of life," a phrase which carries the implication that there is something especially satisfactory, at least spiritually, in farming in a primitive way as our ancestors did upon the frontier. The truth is that "farming as a way of life" is infinitely more pleasurable and satisfactory and profitable when it is planned, scientific, specialized, mechanized and stripped of the long hours and the drudgery of the old-fashioned obsolete pattern of the frontier or general farm.

There is also much loose thinking about "the family-sized farm" and the idea that a specialized, scientific, business-like farm cannot be "family-sized." This is sheer nonsense, for the modern farm may be family-sized and still infinitely more satisfactory than the "family-sized" farm of the past in which there was no program, no plan, no pattern but only a scurrying, planless confusion in which a family was trying to raise a few dairy cattle, a few beef cattle, a few hogs, a few sheep, a few chickens together with 10 acres of this and 10 acres of that in a frontier pattern which no longer has any justification in a highly industrialized world where markets, distribution, mechanization and many other factors which did not exist on the frontier have altered not only the agricultural but the economic and even the sociological picture.

The specialized, scientific, business-like farm does not mean the single-crop or even the undiversified farm. It merely means that a farmer and his family do two or three jobs well on a planned, sensible basis rather than a dozen jobs planlessly, badly or inefficiently. Such a pattern approaches that of the efficiency by which American industry produces more and better automobiles, plumbing, radios, etc., than all the rest of the world put together, sells them more cheaply to the consumer than any other nation in the world, and at the same time pays its industrial workers wages from 30 to

90 per cent higher than any other country in the world. The same pattern of planning and high production per unit, per man-hour, per dollar invested, applies to agriculture as well as to industry, but in terms of high and efficient production per acre, per man-hour, per dollar invested. And, of course, the base is always the quality and productivity of the soil and the production per acre. If the raw material represented by grain and forage is costly because the farmer spends too many man-hours, too much fertilizer and too much gasoline in relation to his yields per acre, it does not matter how many animals he owns as machines in the barns or on the pastures, he will make no money, any more than a factory with hundreds of machines can make money by purchasing raw materials at a price so high that the price for the finished product is completely out of line with his costs.

In the end such a process simply means failure and bankruptcy. Ruin approaches the farmer, whether he is in cash crops or in livestock, at an exact ratio with the decline in yield per acre because as the yield declines the costs go up. As my partner, Bob Huge, states it so well, "Too many farmers think they make their money on livestock or on the *number* of livestock they carry. The animals are merely the machines which process the raw materials. The amount of money the farmer makes is determined almost entirely by the amount he raises per acre."

The livestock industry in Texas is still subject to this ancient fallacy, for too many bankers and too many ranchers estimate the wealth of a rancher by the *number* of livestock rather than the quality and the carrying-capacity per acre of his range pasture. A 10,000-acre ranch running one steer to every 30 acres of thin, overgrazed pasture cannot make as much money as a 5000-acre ranch running one steer to 5 acres of good *undergrazed* pasture. Nor can the corn farmer raising 20 bushels to the acre compete with the corn farmer raising 100 bushels per acre because it costs the 20-bushel-an-acre farmer five times as much in taxes, interest, gasoline, man-hours and seed to raise a bushel of corn as it does the 100-bushel-an-acre farmer. The livestock have nothing to do with it except to process the raw material.

The fact of this changed pattern in American agriculture is one of the things rarely impressed on the farmer in all the operation of our ponderous machinery for agricultural education. The specialized, scientific, business-like operation has nothing to do with the *size* of a farm. The profits, prosperity and low costs of operation are determined by the program and the pattern, and the New Agriculture is not the enemy of the "family-sized" farm but on the contrary its greatest friend. As two outstanding examples, I have one friend who last year grossed $145,000 on 13 acres. He is a truck gardener and a hothouse grower. I have another friend who grossed $44,000 on 160 acres but he did it by specialization and high production per acre. As a general farmer on the same acreage, even in a time of high prices, his gross could not have exceeded $8000, in addition to long workdays and man-hours spent in running about doing a dozen different things with indifferent skill and efficiency.

This revolutionary change in American agriculture, away from the frontier to the highly complicated age in which we live—a change which has been going on unnoticed for the past generation or less —is one of the factors, perhaps the key one, in a modern, successful and profitable agriculture; yet our ponderous educational system has largely failed to bring this fact to the consciousness of the average American farmer. The pattern of the New Agriculture—imposed upon the face of the nation—could vastly reduce the necessity for subsidies, parities, floor prices and other semi-political dodgings of the issue, permit agriculture to stand squarely upon its own feet, reduce taxes and living costs and greatly increase the farmer's income.

One factor which the writer finds peculiarly stupid and irritating is the condescension toward the farmer practiced throughout many of the bureaus and agencies both of the federal and the state governments. It is based, I think, upon the assumption that the average farmer, because he sometimes lacks education, is therefore stupid and incapable of comprehending the simplest scientific fact.

There are many elements in our past agricultural history which tend to lead the educational forces generally into this error. Some of them are bald facts, rarely discussed even by the historian and never by the politician. Among them is the fact that the early pioneers were by no means all splendid, heroic, courageous and intelligent characters as we are led to believe by most history books and romantic novels. Among them, perhaps in the majority, was a large percentage of the shiftless populations of the East, unable to hold their own or stand upon their own feet in the more severe competition of already heavily settled areas. Many of them were bankrupt and a few were actually running from the sheriff. Some were merely adventurers who created nothing in the new world to which they migrated but merely lived off the community.

The race of pioneers was by no means made up solely of the cream

[253]

of the Atlantic seaboard communities and, indeed, as the country further and further west was opened up, the quality of the pioneer actually declined. Much of the lawlessness of the West arose wholly from the character of the emigrants. Largely speaking, the descendants of this element found themselves on the land and in many cases upon poor land which they had not the initiative and the energy to leave. Even today it is this element, isolated upon marginal or wholly non-agricultural land and in certain pockets and backwaters, which is not only our greatest agricultural problem but our greatest sociological one as well.

The failure and the gradual disintegration of this element and their descendants in intelligence, in character, in initiative and even in genes is compounded of many things—of originally inferior stock, of poor nutrition, or miserably low living standards, of exploitation at the hands of absentee landowners. Their cure, economically or sociologically, is not an easy task nor can it be accomplished rapidly, and it will not be accomplished by the crocodile tears of the "liberal" and the sentimentalist but only by better nutrition, better farming methods, better income and in some cases by wholesale migration off the marginal land upon which their ancestors, for one reason or another, unwisely chose to settle.

By no means were all the original pioneers of this category, and, inevitably, as history has proven again and again, the cream rose to the top in the opening-up and gradual development of the vast new areas of the nation. Fortunately the shiftless and lawless elements were eventually subdued, sometimes by the costly system of vigilante committees, and the nation grew and education and civilization advanced. But by the middle of the nineteenth century a new deteriorating process began in agricultural areas which continued steadily until very recently.

It was a process closely allied to the vast industrial and metropolitan development which took place from the early nineteenth century onward. Put very simply, it was no more than the slow but insistent draining away of the best stock from the farms into the great cities where opportunities for educational and material advancement were much greater. It was a simple and an understandable process in which, in most rural areas, the more intelligent, the more energetic, the more creative drifted away generation after generation from the farm into the city leaving behind the less vigorous, gifted and talented to carry on the agriculture of the nation. It was not only that this general migration was disastrous in the immediate sense; it was also cumulative, for those left behind provided, generation after generation, an agricultural population gradually but persistently declining in the qualities of energy, intelligence and ability.

Very simply it worked thus—that in a family of five children, the four most able ones migrated to the city to make their fortunes, to become presidents of banks and railroads. These left behind the fifth and least energetic and clever member of the family to marry in turn a boy or a girl who alike had been left behind. In turn they had, let us say, five children of which the four most able in turn migrated to the city, leaving behind the least able one who in turn married the residual fifth of some other family.

This process has been going on over most of the United States for five to six generations or more, and one has only to glance through Who's Who to discover that the vast majority of our leaders in all fields have not come from the cities but were the brightest members of farm families who migrated into the cities. It is, of course, a process which has been costly to American agriculture and one which no livestock breeder would dream of following in his breeding program.

The most tragic element is that this migration of the best stock away from the farm was almost in inverse ratio to the soil fertility of given areas and to the standard of living. The poorer the area, the more complete was the migration of the better element of the stock. On better soils where living was easier, the income greater and the possibility of making a good living more favorable, much of the better stock remained. Also, tragically, the need for energetic and intelligent farmers was and is greatest in the areas least favored in terms of productive and good soils. A bad or stupid farmer can survive and prosper longer on forty-foot-deep alluvial soil in the Mississippi Delta or in the black soil belts of Texas and Iowa than on the thinner, less well-balanced soils of other parts of the country. Frequently his longer prosperity does not mean that he was or is a better farmer but merely that he had better luck in the area upon which his ancestors chose to settle.

All of these elements tended to produce, together with lack of transportation and communication, the typical "hick" type which was the butt of so many jokes in stage, vaudeville and movie theater until very recently. It is significant that in the talking picture of today the "hick" type is rarely seen save as the dreadful example of a reactionary created by a bad agriculture. The usual farmer of today in the contemporary movie, and especially the young farm boy or

[255]

girl, is intelligent, healthy, active and smarter than his puny, neurotic, city contemporaries. This is a great step forward and it leads to the speculation that perhaps, from now on out, under a good agriculture, the four cleverest, most energetic children of a family will stay on the land practicing the New Agriculture while the "dumbest" goes to the city to become a bank or public-utility-company president. There is even much evidence on both sides to indicate that this process is already under way.

Unfortunately this conception of the farmer as a "hick" was not confined to the theater in all its manifestations but apparently prevailed with the general public and to a surprising extent within the higher circles of our agricultural education. I would qualify the term "higher education" by defining it as the circles in which there are the most college degrees.

It is one of the failures of our fundamental American philosophy that we confuse education and intelligence as much as we confuse plumbing and civilization. One ounce of intelligence is worth a pound of education, for where there is intelligence, education will advance and follow on its own, but where education alone exists, the results can be terrifying beyond even the realms of untutored stupidity. Too many of our Phi Beta Kappas turn out to be postmen or clerks or find at middle-age resources largely confined to the comics and the pinball machine. It is the element in our agricultural educational machine which is educated without being intelligent which has taken the attitude of condescension toward the farmer, best represented by the patronizing pat on the shoulder while giving "good advice" by telling the farmer to do something because it will be good for him rather than telling how and why a certain process or practice works, stimulating his imagination and "sparking" his interest and curiosity so that he will take care of himself from there on out.

Most of our bulletins, most of the speeches made to farmers by government agents are concerned simply with giving him advice without making the least concession to his intelligence or capacity for understanding scientific facts. In the writer's own wide experience among average farmers, he has acquired the greatest respect for their intelligence, understanding and actual hunger to be "sparked." Certainly among the top-notch farmers he has found an intelligence, a curiosity, a capacity for observation and deduction, and even for scientific reasoning far above those of many a professor with a whole string of letters behind his name. Very often the farmer shows a lack of interest in a speech or a pamphlet or a lack of response to them

[256]

because he already knows the information being presented to him in a childishly simple way or because the presentation is simply so damned dull. Nothing is worse than the "ready-made" speech, so common in this era of bureaucracy, which is prepared for the "important" speaker by some anonymous underling, and which the speaker has never seen until he rises to read it in a dull and uninterested voice. And nothing is more insulting to the intelligence of an audience. But it is what we get in greater and greater doses out of Washington and even from many of our college deans and presidents.

All too often the farmer is told to lime his soil without being told the faintest thing about what the effects of liming are beyond the fact that if he limes he can raise legumes. He is not told of its effect upon the ionization processes related to iron and aluminum or the wonderful catalytic properties of limestone. He is considered too stupid and uneducated to understand about trace elements, even if the speaker or pamphlet writer has himself ever heard of such things. He is treated, all too often, as if his only function was to take the advice given him without question and simply go ahead gratefully and do as he is told.

To be sure, the exceptional farmer, the bright boy or girl, the inherently good farmer, can get all the information they want and nothing will keep them from getting it. The bureaus, the divisions, the colleges all supply information, and the average agricultural magazine does an excellent job in true teaching, not simply in advising the farmer, but explaining to him the processes which take place and the reasons for the advice. In theory, there is no reason today why a bad or an unprosperous farmer should exist anywhere in the United States. Information is available on every side, most of it sound and much of it fairly up-to-date, but all too often the farmer has to go and hunt for it, and not all farmers are exceptional and not all farm boys and girls are brilliant. The average farmer is average, and as my grandfather observed, "The average ones are the ones who are important, because there are so many of them." They cannot always go and get it, sometimes because they are too busy, sometimes because they become bewildered by the overspecialization which they encounter at experiment stations and colleges, sometimes because, in fairly remote areas, the facilities are simply not at hand. Sometimes a good farm program or sound agricultural methods have to be sold and there are not enough real salesmen, or the county agent, whose job the selling is, has become swamped by red tape and forms to be filled out.

As one of the countless farmers who visit Malabar put it, "We like to visit successful farms. If we go to the college experiment station we have to go separately to the agronomist, the sheep or cattle or dairy or poultry man, the fertilizer man or a half dozen other specialists to get what we want to know. Sometimes it would take us a week or more and we haven't got the time. Nobody puts it into a working pattern for us. And an experiment plot is not the same as a farm. What works on a prepared plot does not work in a field with two or three kinds of soil. At a successful, well-managed farm we can get an immense amount of information and a lot of good ideas in just a few hours."

His remark implied no derogation of the experiment stations which were set up originally to do research and in most cases do an excellent job. Somewhere along the line there is a missing element in our education and I suspect it is the pilot farm, operated by a legitimate farmer himself in terms of the average farm, under the advice which he gets from colleges and in co-operation with them, where visiting farmers may come and go over the place field by field and animal by animal, knowing that the whole place is not operated on taxpayers' money but profitably on the farmer's own income and capital. It would be a place where he could rub shoulders with the same concrete problems he faces at home, a place where he could examine the texture of the soil and the health and vigor of the livestock and look over the fence and know he is seeing a productive and profitable field and that he can find out how it was made that way. We should remember that every farmer *believes* what he *sees*. It might be that a state college should call for volunteers among the good farmers of each county and let it be known that the volunteers were working with the college and getting *all* the answers and reasons and not merely being told to do this or that.

Perhaps the best solution of all would be a great Foundation, privately endowed, which would not concern itself with research but with collecting and co-ordinating all the information and the results of research which all too frequently become lost or duplicated over and over again at a great waste of energy, time and money, because there is no central agency to co-ordinate them and fit them into a pattern. It would be a center for the creation and dissemination of readable and entertaining booklets and for the distribution of the almost endless supply of good films, some of them made by the government and some by foundations and industrial corporations, which today are all too rarely seen by farmers and least of all by the farmers who need them most. It could well be established somewhere in the Great Mississippi Basin as a center where two or three farms were in actual profitable operation on different programs and where the farmer could come to find out all about the latest machinery. And possibly it could set up a series of regional pilot farms in different areas, supported not by taxpayers' money but operating independently and profitably, and these in turn could become smaller centers of information and education. It would be a national center for the dissemination of agricultural information and possibly one day even an international center for a world in desperate need of just such an institution. It should be a clearing house for all agricultural and economic knowledge and maintain a close, intimate and respectful contact with the farms of the nation, and it should recognize and establish recognition of agriculture as what it is—not merely a way of making a living or even a "way of life" but one of the most difficult, dignified and honored of professions. We shall never get such an attitude or accomplish wholly the great things which need to be accomplished from or by government institutions alone because of the many reasons set forth earlier in this chapter. The synthesis of all that is best can only be accomplished through a well-endowed, well-staffed and comprehensive Foundation, independent of academic restraints and prejudices, politics of every kind and free, above all, of the paralysis of bureaucracy and red tape which gets between the farmer and what he wants to learn.

Certainly such a Foundation, devoted to the welfare of all mankind, would be the greatest monument any rich man could leave behind him. It would be a monument that would carry his name down the ages.

Much of this chapter has been rude and even brutal, and undoubtedly the writer has stepped on as many toes as an elephant at a church supper, but he believes that much, at least, of what he has written is with few qualifications true and has long needed saying. The chapter was born out of a single remark which the writer has heard repeated dozens of times from the thousands of farmers who come to Malabar and remain talking and talking after a long day until the milking is done, the cows are turned out and the moon comes up over the lower pasture. When at last they turn to leave they ask, time after time, a simple but extraordinary question. It is: "Why didn't anyone tell us any of this?" They do not mean: "Why didn't anyone tell us to lime?" or "Why didn't anyone tell us the value of green or barnyard manure?" They have been told all that

[259]

time and again. What they mean is: "Why didn't anyone tell us what goes on in the soil and what it means to our livestock and ourselves?" They really mean: "Why didn't anyone tell us how fascinating is this profession in which we find ourselves?" They are simply hungry for the spark which they are not getting. That spark, my friends, is the essence of good teaching. It is an essence of which we have, for all our $3,000,000,000 in expenditure on agricultural aid and education, far too little.

CHAPTER XIII

Notes to Those Who Make Farm Machinery

The earth which yields us food for our sustenance holds powers beyond our control, for it is in league with the elements. . . . Here for one moment we can see where time ends and eternity begins, and wars become a momentary struggle and fortunes a mockery.

—Clare Leighton *in* Give Us This Day

XIII. *Notes to Those Who Make Farm Machinery*

THIS chapter is largely one of suggestion to the manufacturers of farm machinery in relation to the obligations which constantly increasing knowledge regarding soil, livestock, health, nutrition and agriculture in general is placing upon them. They are suggestions made in the friendliest of fashions and with the best of motives, but they are made from the point of view of the good farmer rather than that of the stockholder or the executive of the big machinery companies.

We hear a great deal about the benefits which mechanization has brought to agriculture and the benefits are undoubtedly great, although in the end mechanization alone does not necessarily improve agriculture or even yields. Such improvement in the end can only be achieved by the farmer himself, and mechanization in the hands of a bad farmer can in reality serve simply to destroy farms and soil more rapidly than the bad farmer was able to accomplish the same result in an earlier period.

During the period of farm mechanization, extraordinary things have happened to the world. Since the first binder and reaper was put into action in the field by Cyrus McCormick (an event which might properly be described as the beginning of mechanization), we have seen the invention and development of the automobile, the airplane, the radio, television and the widespread expansion of the telephone and the telegraph and countless other products of man's ingenuity. The advance of agricultural machinery has largely paralleled this progress of man's industrial and mechanical development, but in comparison to other fields (such as those, for example, of machine tools and the automobile), farm machinery, with the exception of the tractor itself, has made uneven progress and in many ways has lagged far behind development elsewhere. It might be put in this way—that while the mechanization of horse-drawn machinery made advances, the full development of all the immense variety of improvement, made possible through the internal combustion engine and many other non-agricultural factors, has lagged well behind. We have plenty of mechanization in farm machinery which represents no more than the hooking-up of this or that piece of horse-drawn equipment to a tractor and calling it mechanization. Mechanization it may be, in a primitive sense, but it is not *modern* mechanization. The usual manure spreader, the usual mower, the usual side-delivery rake are not modern mechanization at all but merely horse-drawn equipment attached to a tractor, and as such they are about as related to truly *modern* agricultural machinery equipment as the Model-T Ford is related to the Lincoln Zephyr.

It is true that, here and there during the past generation or two, the agricultural machinery manufacturers have produced an outstanding combine or an outstanding field harvester or cotton picker, and for this they deserve great credit, but on the whole agricultural machinery needs bringing up-to-date in terms of the radio-airplane age in which we live. It needs to be made of better materials, with more emphasis placed on durability than upon making high profits out of endlessly selling parts made of poor material and workmanship to replace parts of the same kind which broke in the beginning. It must provide new types of machinery needed by the New Agriculture, so different from the old "pretty" farming, frontier school. It needs more fitting tools which will do two or three or four jobs rather than the process of constantly adding some new tool every time a new agricultural practice gains widespread use. (In the past the agricultural machinery lists have grown steadily year after year, providing more and more tools for the farmer rather than reducing the number of tools he must purchase by combining many uses in one or two intelligently designed and durably built tools capable of a wide range of functions.) This factor is increasingly important, especially for the young man starting out in life who finds the cost of the machinery he must buy virtually prohibitive. And all too often he is paying a high price for machinery of antiquated and obsolete design made of flimsy material. (I speak out of experience, not as a theorist but as a worker in the fields who has had first-hand experience with the poor designing, the cheap and flimsy materials and the general obsolescence of expensive farm machinery put out by some of our biggest companies.) The sale of parts in the farm machinery business runs from 25 per cent upward of the total business, a shockingly high rate. Some of it undoubtedly comes from the rough usage to which all farm machinery is subjected and some from the ignorance and carelessness of the farmer himself, but replacement parts costs also arise from the use of poor materials, poor designing or the improper use of materials in manufacture.

In the past and to a considerable extent in the present the manufacturer of agricultural machinery has had the excuse that he must manufacture machinery cheaply so that the price can be within the range of the average farmer. Each year that excuse becomes less valid and in the not too distant future I doubt that it will be valid at all, for as the economic pressure forces the old-fashioned subsistence farmer out of existence, as the younger generation of properly educated farm youngsters begin to make their influence felt and as farm incomes increase as the level of agriculture rises, the farmer will want farm machinery as good and as modern as the automobile he buys, and he will be able to pay for it. The modern good farmer does not leave his machinery out of doors the year round and, indeed, the good farmer never did at any period. He will be willing to pay a fair price with a decent profit to the manufacturer for good machinery, but I doubt that he will submit to having his pocket picked for poor values at high prices as has happened all too often in the past. Some of the machinery today, with its careless tooling, its old sprockets and chain drives, its infinite variety of nuts, bolts and angle irons, resembles more a drawing of a contraption by Rube Goldberg than a piece of modern farm machinery. The farmer has a right to streamlining as much as any other element of the population.

In the past the whole of the agricultural machinery business suffered from what might have been called a tacit monopoly under which competition was reduced to a minimum and no company made any particular effort to step on the toes of another by producing newer or more popular models of machinery or even better machinery; it was to a great extent a question of the farmer taking what was sold him and liking it. There was little difference in the product of one company from another and little advantage for the farmer in shopping around. All too often it was true that one product was not as good as another but rather as bad as another.[1]

[1] Indeed it was this situation which contributed considerably to the establishment by farmers of co-operatives set up in self-defense. The poor qualities of fertilizers and their high costs in the past was another contributing factor. Co-operatives are not set up by groups of individuals out of pleasure or amusement but because the conditions which bring them about have become very nearly unendurable. All successful co-operatives have come into existence simply because the forces which oppose them most asked for them. Some of the big Farm Bureau co-operatives in some of the states have grown into vast businesses and unfortunately some of them are acquiring the very vices of big business which they were set up to avoid. In other words, they are becoming more interested in "business" and money-making than in agriculture and the welfare of the farmer in general.

Despite the difficulties of establishing a new company or even the manufacture and sale of a new implement—and the difficulties are immense because it means the setting up of a whole national system of distribution and servicing—some new and smaller companies have been established recently with considerable success, arising most of all from the fact that they provided in most cases implements which represented *really* modern mechanization and sturdy and honest construction with good and heavy materials. I think that, in fairness, it can be said that it is very often these smaller and newer companies which have contributed most to real advances in the manufacture of farm machinery in general, as well as to a new and better agriculture, a field to which the whole of the agricultural machinery business may make great contributions if it chooses.[2]

I doubt that today we should have the hydraulic lifts on tractors which have so greatly reduced farm drudgery and increased general farming efficiency if it had not been for a new and smaller company coming into the business and forcing it upon the other companies. The Graham-Hoeme plow and the principle behind it could have been developed at any time during the past fifteen years to the benefit of any company showing initiative, and to the vast benefit, as it has since been proven, of the whole of the Great Plains agriculture and agriculture in other parts of the United States as well. The Graham

[2] It is impossible not to mention in passing the very great contributions made to modern farm machinery and even to modern agriculture by Harry Ferguson, an inventor and manufacturer from the North of Ireland. I mention him not as a manufacturer but as an inventor and to some extent certainly as a genius who has been able to see agriculture in national and world terms and to forsee steadily the advances and needs of a really modern agriculture. Through him came into being the small powerful tractor, so immensely valuable to all smaller farms and to many operations on large acreages, the hydraulic lift on farm tractors, the advance of tillage principles over those of the mouldboard plow and the all-steel construction of farm machinery on a sturdy and durable basis. He is also responsible for a really modern new mower and power-take-off hayrake capable of being operated at road speed while handling heavy grasses and legumes. He has contributed enormously to the principle of consolidating farm machinery so that one implement may be put to many uses instead of having to have a different implement for every operation. Undoubtedly Mr. Ferguson's genius and zeal has improved the quality of farm machinery throughout the whole of the field.

Great credit should also be given International Harvester Company for its co-operation with many agricultural projects and organizations and its direct work with farmers. Arnold Yerkes, working for the company in these fields, has made a notable personal contribution in the field of the nutrition of plants, animals and people in which the company has steadily shown an increasing interest. The same company is responsible for two excellent books written by Karl B. Mickey, whose untimely death in 1948 was a great loss to American agriculture. They are called *Man and the Soil* and *Health from the Ground Up*.

plow, a really modern tool designed for modern rather than a "pretty" frontier agriculture, had to be developed, manufactured and marketed virtually single-handed by a man called Bill Graham at Amarillo, Texas, who refused to be swallowed up and very nearly starved before his company finally survived. Intelligent farmers responded and reacted by buying the plow wholesale once its existence became known, because it was the tool that was always needed but which no company made for them.[3] They had had simply to accept what was offered them although the agricultural principles and the construction behind much of the existing machinery was already obsolete.

Grass farming, which implies heavy yields of legumes and grasses three or four times as great as those of the old timothy and red clover yields of the four-year rotation farm, has been spreading across the country like a prairie fire, but little effort has been made among most companies to produce machinery capable of handling and curing these heavy crops. Most of the mowers and the side-delivery rakes are still in the horse-and-buggy stage. Again it was a new and smaller company which put on the market a really modern mower which, instead of clogging, walks through the heavy grasses and legumes like a knife through butter, and a hayrake that can be driven at road speed without clogging up, and leaves the heavy hay loosely piled so that it can be easily cured instead of manufacturing great heavy ropes of hay which can never be properly dried out.

The greatest cost to the farmer of breakage lies not in repurchase of the broken part but in the time lost in the field and the fact that, while waiting for the part to arrive, the weather may change and he may lose hundreds or thousands of dollars because he has been delayed in getting his crop into the ground or in harvesting it. Therefore the question of the quality and sturdiness of materials going into farm machinery is of the utmost importance to the farmer and should be to the manufacturer.

At Malabar it has been our practice to purchase the individual piece of machinery, regardless of make, which from study and experience has proved to us to be the best on the market and the best suited to our uses. Many units of farm machinery are sent to us every year for testing and for experiment to find new uses for them, and our experience in the field has been a very extensive one, far beyond the range of the average farmer. In our experience with much of this machinery I have often wished that the big executives of some of the

---

[3] At about the same time Ferguson brought out a smaller tool requiring less power but operating upon the same "ripping" principle.

machinery companies had been in the field with me when breakages occurred rather than at stockholders' meetings or acting as "big shots" before some Chamber of Commerce meeting. They could have seen the sixteenth of an inch unwelded tubing or the cast-iron tongues or frames, on expensive pieces of heavy machinery, buckling and snapping because the manufacturer had saved maybe 75 cents, maybe $2 on a piece of machinery costing a couple of thousand dollars or more, thus ruining perhaps two or three hours' or two or three days' work and costing the farmer plenty of money.

And sometimes I wish that I had in the field some of the engineers who worked out in theory a piece of machinery on a draughting board and then tested it on perfectly flat ground in sand or silt loam and then reported that they had designed a perfect machine. I wish they had to deal with some of their lopsided equipment on a piece of rolling ground or watch the machinery snap and break when really put to a test on heavy clay or stony soil. Gray or cast iron is a valuable material for many uses and even indispensable for others, but its place is not in the parts of machinery subject to great stress and strain. Cast or gray iron may cost less than good steel and the good steel may not create the steady market for new parts that gray or cast iron will provide when wrongly used either through stupidity or false economy, but somewhere along the line the manufacturer, it seems to me, is obliged to give the farmer a break. All too often both executives and engineers live and move in a world far removed from the problems the average farmer encounters in the field. The nearest many of them ever come to it is at the State Fair where a piece of machinery is whirling about prettily in thin air or where sand is flowing beautifully through a drill or corn planter which would clog up if ordinary fertilizer were used.

The truth is, I think, that the New Agriculture is well out in front of most of the farm machinery being made today and that on the whole the best contributions are being made not by the biggest companies but by smaller and newer companies which are making fewer implements but implements adapted to multiple use so that the farmer need not sink himself economically by buying a wide variety of tools each designed for a single use. It is as a rule the smaller companies which have introduced the hydraulic lift, the power-take-off manure spreader, the Graham plow, the offset disk, the disk plow, the Seaman tiller, the power-take-off top-dressing spreader for fertilizing pastures and meadows, and it is the smaller companies which very largely have contributed a sturdier construction of better materials which have reduced the farmer's repair and parts bills and saved him and the nation as a whole hundreds of thousands of dollars a year in losses caused by delays in seeding or harvesting a crop.

I do not believe that anything written here is unfair and I am certain that any farm machinery manufacturer examining the situation honestly would admit that there is much truth in all the statements made above. I am certain that every good farmer would agree, some of them perhaps with almost too much heartiness. Fortunately the situation is changing and great improvements are being made each year both in modernizing machinery and in better and sounder construction, but there is still a considerable distance to go before agricultural machinery emerges wholly from the horse-and-buggy age and catches up with the New Agriculture and the agriculture of the future. The manufacturer of farm machinery has a great contribution to make by providing the new machinery necessary for the progress in production and profit which is the principal potential factor of an agriculture in the process of rapid change.

CHAPTER XIV

---

*The World Can Feed Itself
If It Wants To*

Since Creation men have joined to conquer Nature or separated to fight for her fruits. Science has furnished them increasingly effective tools to make Nature more productive and increasingly effective weapons for seizing the larger share of the goods produced.

—SHERMAN JOHNSON, *Farm Science and Citizens,* "The Year Book of Agriculture" (1943)

XIV. *The World Can Feed Itself*
If It Wants To[1]

AFTER more than a hundred years the ghost of Thomas Robert
Malthus is walking again. You may remember that it was
Malthus who wrote a pamphlet about the end of the eighteenth
century setting forth the theory that while food supplies of the world
increased by mathematical progression population did so by geo-
metric progression, and that the day was certain to arrive when the
population of the world would exceed its capacity to feed itself. He
also held that populations of any new country inevitably increased
until they arrived at the limit of food production.

Since Malthus died, much has happened to justify the Malthusian
theory in a world where half the population suffers from malnutrition
and the diseases arising from it and where at least two hundred
millions of people live and die without ever having had enough to
eat one day of their lives. Malthus held that the only factors which
served to keep populations in control were wars, disasters, famines
and vice. What he did not foresee was the degree to which all these
factors have been nullified and, living at the beginning of the Industrial Revolution, he did not foresee that, through industrial processing, many nations like England, Germany and Japan would in time
increase their populations far beyond the power to feed themselves
and depend upon industrial profits to buy from outside their borders
the food that was required. Nor could he envisage the fact that,
despite the slaughter of whole political and racial groups, the devastation of cities and millions of dead in the armed forces, the population
of Western Europe would increase nearly twenty millions during
the period of the Second World War. Nor did he foresee the stupendous advances in medicine, sanitation and nutrition which have
increased the life span of the average citizen by at least a third, thus
keeping alive to be fed many millions of people who in Malthus' time
would have died at middle age or earlier and adding still another

[1] The material of this chapter was originally used for distribution by the
United Nations Educational and Cultural Organization and at the 1949 Conference of United Nations Food and Agricultural Organization.

great segment of population to the already great burden placed upon the agriculture of the world. And of course he failed to foresee the advances made in medicine which have kept alive great numbers of the weak and handicapped which in his day would have perished, and which permit them in turn to breed and still further increase populations. The anti-biotics, which have to a large degree nullified the ravages of vice by curing venereal diseases which frequently killed great numbers of the population in Malthus' time and rendered countless more sterile, had not been dreamed of. Nor did he foresee the great advances in transportation and distribution which have done away in a large degree with wholesale famine or so softened its rigors in some areas, as in Bengal in 1944, that, as one Indian put it, "The famine was a failure because it did not kill off enough people."

In Malthus' time whole vast areas of the colonial world were opening up and the prospect of a world shortage of food seemed preposterous to many of Malthus' contemporaries, but now we know that, as Malthus predicted, in most cases these areas filled up almost at once to the limits of their capacity to feed their populations under existing agricultural and horticultural practices. Moreover, many of the newly opened areas, instead of becoming treasure houses of food, turned out to be unsuited to productive agriculture because of the soils or climates, and many of them, because of excessive rainfall and its leaching effect, produced food only of an unbalanced and deficient nutritional value.

All in all, time and the progress made by man himself as well as the ignorance and greed of mankind have served to vindicate Malthus and produce a world in which the problem of food shortages becomes one of an increasing menace. The whole school of Neo-Malthusians have been superficially right. Before their eyes exists the spectre of a world with steadily increasing populations in which productive agricultural land is constantly being destroyed by wind and water erosion and a wretched agriculture in which the yields per acre have been constantly going downhill. Of course when the graphs of these two progressions, one upward, one downward, cross each other, only disaster and an increasing disaster can result.

Opposed to the Neo-Malthusians there is a less pessimistic group of experts which believes that the world can feed itself if it chooses to change its ways and if peoples and governments discover that food is not only vital to the world but that it plays a great part in the question of war and peace. In the past the writer has loudly cried "woe!" and "havoc!" over the destruction of the natural resources not only of this country but of the world. This shock treatment is immensely valuable in arousing the torpid, the complacent and the ignorant to the menace of this destruction. I doubt, however, that it should be carried to the point of creating the kind of despair that brings lethargy in its train; and that is what the Neo-Malthusians may possibly accomplish. Beyond these schools there exists a whole school of so-called experts which has, often through the medium of newspapers and national magazines, been conducting a sort of "Pippa Passes" campaign that "God's in his Heaven and all's right with the world" and there is nothing for us to worry about. This silliness is inspired sometimes by ignorance and in some cases at least by bureaucratic and academic feuds and jealousies. As a practical working farmer I am inclined to believe that the answer lies somewhere in between the two schools. The world is not feeding itself and actually is on the downgrade but it could feed itself if it chose.

I am not optimistic regarding the prospects of peoples and nations changing their ways to correct a tragic situation; the record of history stands against such a change since mankind rarely if ever undertakes a change or a reform until circumstances become so grim that there is no other way out. Yet in this country at least some great changes, indeed astonishing ones, have come about within the past few years in the whole field of agriculture and horticulture, changes and advances and discoveries so great indeed that they point the way very clearly to the likelihood that the world could feed itself and feed itself perhaps better than it has ever been fed if it chose to do so. No nation in the world has had a worse record than ours in exploiting and destroying its natural resources. Certainly no nation in the world has ever made so much progress in the wrong direction so rapidly. Yet, at the moment, we are beginning to provide what might be called an historically astonishing spectacle—that of a nation and a people doing something about a crisis before it actually becomes a disaster.

Today it could be said, I think, that the world is turning more and more to this country for the answer to the problems of soil erosion, depletion and waning food supplies. Perhaps 50 per cent of these answers have come out of the billions spent upon federal and state agencies, and the remainder through privately established or industrial research foundations and from the better category of farmers—the kind of farmer possessed of intelligence, imagination, curiosity, capacity to learn, and blessed with the powers of observation. The truth is that during the past generation or less there has been a great revolution going on in American agriculture which until quite recently has passed almost unnoticed. It is a revolution which involves many factors—mechanization, technology, economics, plant breeding and wholly new approaches to soil, nutrition and disease and to the creation of new sources of food. Analyzed, the revolution presents an amazingly complex pattern with its threads extending into the fields of chemistry, physics, medicine, nutrition, health and almost every manifestation of science and human endeavor. Perhaps only the first-rate farmer and a very few teachers and experts above the level of the specialist or the fustily academic have comprehended the size and scope of this revolution and what it could mean if its benefits were applied to the agriculture of the world in general. Its possibilities could have an immense effect even upon the establishment of world peace, for food and raw material and markets and purchasing power all lie at the very root of the problem of peace in our shrunken modern world.

I suppose it could be said that modern agriculture began with Liebig and the invention of commercial or chemical fertilizers. His creation of chemical fertilizers revolutionized agriculture, but in the end it created as much harm as good for it gave rise to a whole school of agriculturists who, with the oversimplification which sometimes handicaps specialists, looked upon the chemical fertilizers as *the* solution to the whole problem of producing great quantities of food and fibre, continuously and at a comparatively low cost. The assumption proved to be disastrous. As a result of the use of chemical fertilizer alone, millions of acres of good farmlands declined rapidly in production and some millions were virtually destroyed, thus providing a fine but costly example of the scientific error of concentrating upon one factor to the exclusion of many others. By the persistent and sole use of chemical fertilizers the texture of soils was destroyed, erosion and the attack of diseases and insects of all kinds were promoted and adequate moisture was eliminated together with all those factors such as bacteria, fungi, moulds, earthworms and other countless organisms, some of them as yet undiscovered, which we know now are indispensable to a real agriculture and a sound soil capable of optimum production.

All that has changed with the coming of the agricultural revolution and the access of new knowledge regarding soils and their productivity and the laws which govern these things. Today manufacturers of chemical fertilizer are among the greatest advocates of organic material in soils for, wisely, they have come or been forced to recognize the fact that their product is beneficial or destructive in exact ratio to the amount of organic material, moisture or drainage in the soil, whether abundant or lacking. They have realized that a good and permanent customer is a satisfied customer and that a farmer who understands soil and good agricultural practices is not only satisfied with results of chemical fertilizer but is prosperous enough to purchase it persistently and in quantities profitable to the manufacturer.

This error about the panacea qualities of chemical fertilizer did not come from Liebig himself—he was also considerable of a biologist and a botanist and a generally cultivated man; it came from the lesser men following him who were perpetually looking for panaceas or short cuts in agriculture.

The chemical fertilizer error which began the agricultural revolution has been almost wholly corrected in good agricultural practice and teaching and much of the correction has come from a whole chain of new discoveries which have brought the agriculture picture back into balance, as we have come to learn and understand that a cubic foot of soil capable of optimum production in terms of quality nutrition as well as quantity is a highly complex affair in which can be found in operation virtually every natural law governing the universe.

The point is that if every cubic foot of the world's soil now under cultivation could be made into a perfect, balanced soil and one of optimum production, not only could we raise many times the amount of food the world is now producing but we could improve its actual nutritional quality to such a degree that, save for inherited organic deficiencies or weaknesses, we could alter the whole physique, character and intelligence of millions of the world's population for the better. In other words, instead of producing coolies or "poor whites," living perpetually on the borders of malnutrition and even starvation, we should be producing well-fed, intelligent physical specimens capable of producing and enjoying economic wealth and of actually making continuous contributions to civilization.

The task of converting the agricultural soils of the earth to such a high degree of organic and mineral perfection would be a colossal task but not a wholly impossible one. The means are at hand and known and available to every government and every people if they choose to use them. That they will choose to do so is improbable, but the more we move in the direction of that goal the better off we shall be—you and I in the city apartment or on the farm as well as the coolie in India or China or the tribesman in darkest Africa. Today less than a fraction of 1 per cent of the agricultural land in the world could be included in the category of optimum-production land. That fraction would include a few thousand farms and truck gardens. The rest of the earth's agricultural land is producing only a fraction and frequently a fairly small one of what it could produce.

One needs only an observant and perhaps a somewhat experienced eye to discover the immense disproportion between potential and real production on our American farms, even in the rich Middle West, or to discover the fact that, conservatively speaking, American soils now under cultivation for food production are turning out less than a third or a fourth of the production of which they are capable under a sound agriculture. This fact is evident either from a train or an automobile or an airplane. You have only to look in order to see it.

From the airplane one has a special point of view which reveals, in a trip across the continent, the millions of acres of fields wholly ruined or reduced to low production by erosion. From the air one can also discover, especially at spring plowing time, the millions of acres of land in low production and producing less and less each year simply because the soil structure is constantly deteriorating and the content of organic material is constantly declining. And one can see millions of acres of potentially good productive land growing only broom sedge, poverty grass, golden rod, sumac and other low category weeds. Virtually all of this land could be producing cattle and poultry and dairy products. Indeed, very often these weedy hills are better suited to the production of these essential high-protein foods, so badly needed in the world, than the flatter deeper-soiled areas.

In 1949 I kept watch from a train window for a hundred miles out of Omaha going through Southwest Iowa without seeing one field out of ten in a sea of corn that was producing much more than about 30 bushels to the acre. Under a proper agriculture, the same land, within a few years, could be producing somewhere between 80 to 100 bushels. A few fields, reclaimed under the New Agriculture from erosion and organic depletion, were producing that amount of corn and better. In midsummer of the same year, I traveled from South to North in the rich state of Ohio without seeing more than one farm in ten which was producing more than 50 per cent of its potential production under a proper agriculture.

These observations were made in two of the richest agricultural states.

About 10 per cent of American farmers, the ones practicing a sound agriculture, produce 50 per cent of the food consumed by our population. The remaining 50 per cent of our population is largely fed by about 30 per cent of our farmers who are pretty good at their jobs. The remaining 60 per cent of our farm population produces either cotton, tobacco or similar crops or very little more than it consumes.

One is accustomed to hearing the expression "worn-out farm" to describe farms which have reached a low stage of production or have been abandoned altogether. It is a term which is very likely to become obsolescent and eventually obsolete as we make progress in our understanding of soils and agriculture. The writer is inclined to believe that there is no such thing as a "worn-out farm" except in the case where erosion by wind or water may have carried off all top and subsoils down to the level of unproductive hard pan or actual rock. All soils were originally nothing but accumulations of minerals and rocks in broken-down form. As life progressed and developed on this planet and the process of organic birth, growth, death, decay and rebirth advanced with increasing rapidity, these inorganic mineral accumulations became overlaid with deposits of minerals in the organic form of dead or decaying animals and vegetation and in this organic form the minerals were highly available for the nutrition of succeeding animals and vegetation. When man began the first agriculture, he began destroying this thin organic residue as rapidly as possible, generally speaking in two ways: (1) by encouraging it to wash away down the streams and back into the oceans; (2) by burning up rapidly through cultivation the organic content so that not only did he gradually destroy all the living qualities of the soil (the moisture, bacteria, fungi, moulds and other factors dependent upon decaying organic materials for their existence) but he also destroyed the process by which these organisms operated to break down chemical combinations of inorganic minerals, unlocking them and making them available to plants and consequently to animals and people. In other words, he broke or destroyed completely a natural chain reaction together with the balance and the laws by which soils are created and recreated and by which they can constantly restore themselves and make available to plants, animals and people the minerals existing in inorganic and frequently

[279]

unavailable form and in certain areas in virtually inexhaustible amounts.

On the whole, this destructive process has been called agriculture for thousands of years. Only here and there in very limited regions and among a few farmers throughout history have there existed exceptions to the general pattern.

It seems to me reasonable to suppose that the average "worn-out farm" is not worn out at all. At least, below a depth of the few inches which have been overworked and carelessly farmed, the same combination of minerals exists as always existed, but through the breaking of natural balances and the universal law of birth, growth, death, decay and rebirth and the absence of sufficient moisture and of all the organisms necessary to a *living soil*, these organic minerals have simply become unavailable. Actually they are there all the time, and in most cases all that is necessary to restore fertility is the application of good farming practices and of calcium and of some chemical fertilizer to get the ball rolling and restore the deficiencies of the overworked top few inches. Such a theory is not, of course, wholly applicable to those soils which were unbalanced or deficient in their virgin state but most of these soils have not been put to agricultural use or have quickly demonstrated their deficiencies and have been abandoned. However, there are thousands of "worn-out" farms in some of our richest agricultural areas where the original mineral fertility still exists in the deeper layers and even to some extent in the long-worked upper level. The fertility is simply unavailable because of poor farming methods.

It is one of the common erroneous assumptions of the layman that crops consume great quantities of minerals out of the soil, thus depleting it rapidly. Actually no such thing occurs. No plant, even greedy corn or cotton, takes as much as 5 per cent of its weight or growth from the soil. Many plants take much less than that. The rest comes out of sunlight, air and water through the miraculous process of photosynthesis which ends by providing us with the proteins and oils. That small amount of minerals, however, is necessary in order for the process of photosynthesis to take place and the capacity for production of fibre and seed is in turn determined by the quality of balance or imbalance existing among the minerals in the soils. In many cases, it is not actual mineral depletion which reduces crop yields to the point where a farm becomes "worn out" but that poor agricultural practices have made the minerals unavailable. The same minerals are still in existence but they have been reduced virtually to the same condition in which they existed at the time that organic life began on this planet.

In the hands of a truly modern, intelligent farmer these "worn-out" farms can be restored rapidly and at an economically possible cost to the level of their original production and sometimes—through the addition of fertilizers to correct poor mineral balances and by new, hybrid and more productive plants—may be raised above the level of the original virgin soil production in terms of quality as well as quantity.

This is an operation which I have observed taking place or already achieved on great numbers of so-called "worn-out" farms through the South, the Atlantic seaboard and the Middle West. It is exactly the process we have employed on our own worn-out land at Malabar Farm where today we are growing more of most crops per acre than the same land ever did even in its heyday of rich virgin land production. In the process we have spent less money on fertilizer than is spent annually on many farms providing much lower yields. The process is neither expensive in terms of money nor is it in any sense a short cut; it has been simply that of working with Nature rather than against her, of following the very method by which Nature built up some of our richest soils *but* by speeding up the process immeasurably through the technology, mechanization and knowledge which man has provided and which are now available to the whole of the world. In the same area of once eroded land, no more erosion occurs today with the land under intensive cultivation than when the same land was covered by hardwood forest. Possibly there is even less erosion.

It is to these low-production or "worn-out" farms that we must turn in this country for further increase in agricultural production since we have no more fabulously rich virgin land to be had for the taking to be turned into high production. To a great extent this is also true of the rest of the world.

As has been stated elsewhere, Nature laid down her accumulations of minerals in a haphazard way and in combinations of great variety so that, contrary to the belief of the average person, all virgin soils are not of necessity good or productive soils.

In laying down minerals and in the process of translating these from inorganic to organic form, Nature has created a few soils which were originally very nearly perfect. Among these, perhaps at the very top, are the Black Belt soils of Alabama, Texas and the Ukraine. In Alabama these soils, first depleted of organic materials and converted into "dead" soils, have been allowed to erode away until in the greater part of the area the soils are no longer black but gray. In Texas the same soils put under the plow from seventy-five to one hundred years later have also suffered from erosion to a lesser degree but, while they remain *black*, the yields per acre have fallen off as much as 50 per cent or more during the past few years. This is not because of mineral depletion but because the destruction of the original organic materials and the failure to replenish them has made the Black Belt soil a *dead* soil in which the natural mineral fertility has become unavailable. Soil specialists in the area do not recommend the use of chemical fertilizers because the natural mineral fertility of these soils and the quality of their mineral balance is still so great that chemical fertilizers would add nothing to the potential fertility, but organic material of any kind mixed into the soils produces miraculous results. A crop of Hubam or annual sweet clover plowed into a field can raise production in one year as much as 30 per cent or more, simply by reintroducing into the soil the natural process of birth, growth, death, decay and rebirth which maintains moisture and makes the native minerals available to plants and consequently to animals and people. This is only one of the many instances where simply by proper agricultural methods the yields of soils can be increased enormously, in some areas, similar to our own in Ohio, even increased many hundred per cent.

The great wheat area of the Southwest offers another striking example of the effect upon production of simply applying knowledge and common sense to agriculture. In the past the rule for three years was on an average one total crop failure, one fair year and one good year. There has been *no* failure in the wheat crop since 1939 and in nearly every year the crops have been of bumper proportions. It could be said that the dust storms (and this was the great dustbowl area) have virtually ceased or at least become almost sharply localized in individual farms or small areas where a wretched agriculture is still practiced.

This change was brought about *not* through changing climatic conditions but by a simple technological change in agriculture. The old-fashioned mouldboard plow has been abandoned and a "ripping" plow, which tears up the soil but does not turn it over, burying all rubbish and organic material, has been substituted. No longer do the farmers, who know on which side their bread is buttered, burn over the residue of straw leaving the ground bare. Instead they chop it or work it into the surface sometimes to a depth of four to five

[282]

inches. Some farmers are using subsoilers or great chisels which rip up the soil to a depth as great as twenty inches or more, breaking up the hard pan and admitting to the mineral reserves deep down both the air and the moisture which help to make them available to the crops growing on the surface.

In that area in the past, the success or failure of a wheat crop was virtually determined by the amount and timing of rainfall. Today this element of chance has been virtually eliminated simply through a purely technological change in agricultural methods. Although rainfall has been below normal in the area during the past three years and not too well distributed, there has been no crop failure and in some years the yields have reached bumper proportions. The answer is, of course, that through intelligent methods backed by modern machinery a twelve-inch total rainfall is today as valuable as an eighteen-inch rainfall under the old pattern of agriculture since nearly all rainfall is trapped and insulated. The likelihood of a crop failure or even poor yields has been largely eliminated, and at least one great source of the world's most valuable food, wheat, has been largely insured against the caprices of weather.

In all of these advances, technology and in particular mechanization cannot be overlooked. Under proper land use and with proper consideration for the soil, mechanization has increased yields enormously not only by establishing the possibilities of working soils to better advantage but by cutting to a minimum the crop losses arising from the weather—at planting or cultivating or harvesting time. Today if a farmer has a few bright days he can, through the proper use of mechanization, plant, cultivate or harvest a crop and save it all where under conditions existing in the past under animal power or by slow hand labor he might have lost part or all of his crop. Mechanization has introduced as well a great number of tools such as the subsoiler, the field tiller, the Graham-Hoehme "ripping" plow now in use throughout the Great Plains area, and many other implements adapted to the New Agriculture and designed to achieve optimum production. Moreover, mechanization, together with all the other factors discussed above, can bring down the cost of food, increase the profits of the farmer and increase the purchasing power of the dollars of all of us. What we need is not more and more dollars, higher and higher prices and higher and higher wages but dollars which will buy more. That is the only real raise in wages or income that any of us can get and the New Agriculture offers us the greatest single field in which this can be achieved.

[283]

Meanwhile in other fields and areas ranging from irrigated to dry farming the amount and scope of our knowledge regarding agriculture and nutrition is advancing at a breathless pace. Scarcely a day passes without some new discovery which in turn leads on to still newer discoveries. Agricultural scientists today are making valuable new discoveries through the use of the spectroscope, an instrument which has hitherto been used to determine the mineral contents of distant planets. Today the spectroscope can be used to determine not only the mineral composition of soils down to the smallest amounts of trace elements but it can determine the mineral content of vegetables and so fix their value in the realm of nutrition.

In a half-dozen places agricultural research workers are using radioactive isotopes of various minerals to trace the parts each mineral plays in the growth and productivity of any plant. It is highly probable that within the near future these workers will be able to understand and explain one of the miracles of the universe—how a plant is able by the process of photosynthesis to turn sunlight, air minerals and water into carbohydrates, proteins and oils, and provide us and the whole animal kingdom with the abundant basic means of existence upon earth. The prospects opened up by such a discovery are almost endless.

I am by no means so pessimistic as the Neo-Malthusians and certainly I do not accept with philosophical despair the prospect of half the world starving to death while the remaining half destroys itself in a struggle over the remaining food. I am writing not out of dry figures but out of a life of long practical knowledge of soils, crops and agriculture around a good part of the world and with a knowledge of most of the immense advances made in agriculture and potential food production during the past generation.

The world can certainly feed itself much more efficiently than it is doing today simply by better distribution and the annihilation of the truly vast waste which occurs principally in those nations where good soil and abundance allow a wide surplus over the needs of the nation. It can be accomplished by more efficient farming and a more widespread mechanization designed to increase yields and decrease costs of production. Food supplies can be increased in some nations as much as 300 per cent by applying the knowledge we now have concerning soils and by the use of modern fertilizer programs. Food supplies can be increased by the proper use of land, i.e., growing wheat upon land primarily designed by nature for wheat, grass upon land which is not, for reasons of climate or rainfall or soil deficiencies, too expensive to remedy for the use of grain production; by feeding our livestock population largely upon such soils and decreasing the wasteful and economically extravagant use of grains for the production of animal high proteins, when good grass and legumes can achieve the same results or similar ones at a much lower economic cost and at the same time provide release of the bulk of grains for use directly as cereal proteins by great starving areas of the world. (It should be remembered that in meat or poultry or dairy products, it requires about 7 pounds of high-protein grain to produce about 1 pound of high-protein meat, milk or eggs derived from grain feeding. All of the beef and all of the dairy products can be produced efficiently and in quantity and quality without the feeding of any grain whatever, and the grain rations behind swine and poultry products can be decreased without harm in some areas as much as 70 per cent to release that amount of cereal grain for people to eat directly.)

Various statistics have been given by various authorities regarding the acreage of soil left per person to feed the world. These range from about three acres to under two acres. The statistics apply usually to the amount of soil now under cultivation or at most to the amount of agricultural land available under existing agricultural methods. Both of these estimates, while perhaps valuable as giving a general picture, are very loose and in general relative, for even a shift in agricultural methods and land use could make much land now unused, and formerly considered unusable, into productive, food-producing terrain and alter considerably the average acreage per person of world population. Nothing is more difficult to arrive at than such estimates because they are subject to so many factors. Indeed, it might be said that, under rapidly changing advances, chemical, technological and otherwise, no such estimate is really anything like valid or exact for longer than twenty-four hours.

Using the soil as the basic producer of food, as it is today (but need not necessarily be in the future when the sea, fresh-water tanks, laboratory vats and other mediums may become immense producers of food), we have learned pretty accurately what makes a good productive soil both in terms of quantity and quality production, and we have learned as well how we may, quite literally, manufacture such a soil. It is technically possible, of course, to make good productive soils out of an asphalt pavement or a collection of old ash trays, but it is not always possible in terms of economics because it costs too much. Somewhere between good natural, virgin, productive soil, created for us by Nature herself at no cost, and the other extreme of "manufactured" soils mentioned above, there is a line at which we can restore worn-out soils or create good soils out of naturally deficient ones at a cost which is economically possible.

Even this economic line cannot be determined by the simple process of profit and loss or return on investment because when problems of food become acute, such soils are created or restored and put into production *regardless* of any cost save necessity. Under such a stress the problem becomes political and only secondarily economic, and in nations short of currency exchange and in capacity for feeding themselves, we are already finding such action becoming a political necessity (as to some extent in Great Britain today).

Under such stresses the acreages which *can* be converted into a food-producing medium become almost unlimited as compared to our conception of potential agricultural land in the past.

In the field of poor land use and low production per acre, one found in the United States until quite recently the obvious and extreme examples of both factors in the Southern and Southwestern portions of the nation. Here millions of acres of land are utilized for growing unwanted cotton, a crop for which the demand declines constantly, largely because of technological advances in the field of plastics and synthetics made more cheaply from other sources of cellulose. Other millions of acres are used for the growing of corn or maize, a crop unsuited to much of the area because of excessive heat and light or erratic rainfall. The average yields of corn range from 12 bushels per acre in Georgia to 16 in Oklahoma. Obviously all of this land, whether given over at present to cotton or corn, could by an adjustment of program and the choice of different crops and better seeds (good hybrids rather than low-production white or yellow open pollinated varieties) produce truly immense increases in the amount of food to be delivered to a starving world.

In many respects the whole of the pattern set forth above applies to the agriculture of most nations. At the base of all of this lies the basic truth, so often set forth by Dr. H. H. Bennett of the U.S. Soil Conservation Service, and so frequently repeated in this book, that poor land and poor agriculture make poor people and poor nations. It is a kind of vicious circle with one evil influence operating potently upon the other to increase the pattern of degeneration both in soil and people.

The intensely acute food problems of China and to some extent those of India are created not alone by overpopulation (which is

[286]

always relative) but by the poor production per acre of a declining fertility of soils from which even the dung and the roots of plants are taken to be used as fuel and by the almost total lack of commercial fertilizer to aid and increase production per acre per capita.

In China and until very recently in India the traditional methods of land-holding in small scattered bits and the tax farming system employed by the great landowners were both immense sociological factors in depressing both society itself and the productive capacity of the agricultural land.

It is not at all exaggerated to say that China, after socio-political changes in land tenure and technological and mechanical improvements in agriculture, could produce easily half as much more food as she produces today. In India such changes could bring about easily an increase of a third in food production, the effects of which would in both countries be enormous, not only in the nutritional and economic fields but in the political ones as well.

It is doubtful, however, that either country, with its present population and the population increases recorded (very vaguely and inefficiently during the past few years), could ever, even with our present great advances in agricultural knowledge, feed itself properly. It is conceivable that further advances will make something resembling a balance between food and populations possible, but today Japan and Java alone seem to have achieved the maximum result in bringing the food supplies somewhere near the demands of a population living at very low nutritional standards, and even in these countries, where very little land is wasted and where agriculture is generally on a fairly high level, there is a serious gap which in our times can only be filled by better distribution from other areas where population pressure is less and food surpluses exist. At least one can say that in Japan and Java agriculture is efficient and productive at very nearly the maximum level within the limits of our present agricultural knowledge.

The United States, of course, stands at the top of the nations and areas which are continually plagued not by a shortage of food (except during wars when arbitrarily she exports great amounts of food purely as a war measure or checks production through bureaucratic controls) but by surpluses which glut the market, run down the price of food for the farmer who raises it, and curb his purchasing power which in turn depresses and undermines the economy of the whole nation.

As a rule all sorts of makeshift measures are employed to cut down production or to store the surplus (a high percentage of which either goes to waste or is destroyed by government to keep it off the market). All of these measures designed to limit food production in a starving world are not only silly but even idiotic. By the use of only a fraction of the money paid out in subsidies not to raise crops, in the wastage and actual destruction of food and in the money expended to keep food off the markets in order to maintain prices, means could undoubtedly be worked out, even to the construction of refrigerated ships to carry potatoes, citrus fruits and other perishable and valuable food crops which are habitually destroyed in vast quantities, by which these vast surpluses could be carried to other parts of the earth where people live perpetually at starvation levels. The problem here is not one of production or of a shortage of food but of economic co-operation among nations and just a trifle more intelligent planning on the part of bureaucrats and politicians more concerned with buying local votes than with relieving human suffering, starvation or even establishing world peace. In the United States there is no problem of trying to raise more food, as is the case in many parts of the world. The problem is merely one of distribution, and secondarily of an improved and more productive agriculture at a lower cost of production.

Beyond all of this, in the realm of land use and production there still remain very large regions which can be put into production by terracing, by irrigation or by drainage. Indeed, the whole area of possible agricultural land that is usable or being used on a low production basis is in the United States alone immense. In a large degree this is also true of most nations in the world. And there is finally the element of improved strains of crops and seeds in which our scientists have made great advances. Recent figures available from Italy show that, by the use of American hybrid varieties of the yellow maize so popular in the Italian diet, gains from 20 to 90 per cent to the acre have been achieved over the yields from the old primitive and traditional open pollinated varieties. This means, of course, a potential gain in the food supply raised on the same acreage now under cultivation for maize in Italy of from 20 to 90 per cent. Similar gains from improved and disease-resistant strains of grasses and grains and vegetables can in themselves vastly increase the world supply of food almost everywhere.

Thus far I have been concerned with the food or the potential food that is derived or can be derived from soil. Beyond the soil lies the sea in which undoubtedly there are still vast unutilized reserves of high-protein food. The waters of Chesapeake Bay, a warm and shallow area, produce more high-protein food per acre than any acre of land in the world, but again, under the mismanagement of man, this production has been cut during the past generation or two by as much as 25 per cent through deposits of mud coming from adjacent agricultural land eroding under bad management and by sewage and industrial pollution from the cities. Under good management, it is possible that not only could that 25 per cent loss be restored but that the production of seafood, by planting and other methods, could be enormously increased in the Chesapeake Bay area alone. The United States Government is today conducting, at Woods Hole, Massachusetts, an experiment in growing clams (a highly nutritious high-protein food) by the cropping method of planting and harvesting as one would plant and harvest a crop of wheat.

There are also immense possibilities of producing high-protein food by raising fish as crops, in managed fresh-water ponds, possibilities which have only been touched upon in a few very limited areas of the earth. Any acre of well-managed pond will produce as much high-protein food or more than the acre of land adjoining it, and fish farming can flourish in areas where soils are comparatively poor or even unproductive. Such fish farming has been successful in Austria, in Java and in China, successful not only in producing food but in providing a highly profitable enterprise as well.

In Texas the Southwest Foundation is actually experimenting with the "ranching" of fish in the Gulf of Mexico. In that great body of water the temperature rarely falls below 70 degrees and this temperature favors the high production of the plankton upon which feed in turn the shrimps and many smaller fish and organisms which in turn become the food supply of the larger fish. By the use of electronic action in the water, it is possible to "herd" these fish into concentrations where the crop can easily be harvested and even to "fence" them within given areas and feed them artificially.

Recently a rancher friend of mine in the Brazos River bottom had occasion to drain one of the land-locked bayous on his land, and the drainage revealed a truly immense quantity of fish, ranging from the highly edible bream, perch, bass and catfish to a monstrous gar weighing 120 pounds. The weight ran into several tons, which was trucked and sold at the Houston markets for a sum representing a much greater return per acre in cash and profits than the best acre of his good bottomland returned either in pecans or in cattle production. I pointed out to him that if he put the bayous to work producing fish he could, in the warm climate of the Gulf Coast and with the abundant food of the bayous, harvest every three years a similar tonnage of fish, making the land covered by the waters of the bayou from three to four times as valuable as the dry land given over to pecans and cattle.

In experiments made by the State Agricultural College in Georgia, the use of fertilizer in small ponds increased the yield of the fish crop as much as 500 pounds to the acre.

At the recent food and agricultural conference held at the United Nations at Lake Success, scientists of all sorts brought out information regarding astonishing methods of creating high-protein foods or proteins for the fortification of low-protein foods such as rice which can be produced in vast quantities. These proteins and some fats are produced by bacteria, by yeast, by photosynthesis, even by moulds and fungi out of waste materials ranging from brewers' mash through sawdust and even by algae growing in fresh-water tanks or seaweed rooted in salt water. During the Second World War great advances were made in Germany and Sweden in the production of high proteins from sawdust and other waste products of the timber industry.

At one time it was believed that one of the serious factors in the declining food supplies of the world would be the exhaustion of the elements and minerals which go into the commercial fertilizers that produce not only quantity but quality production of foods. Discovery of new and in some cases almost inexhaustible supplies of these minerals and elements together with the production of nitrogen out of the air itself at low cost for fertilizer have largely allayed these fears. But beyond these discoveries lie other immense sources of fertilizer, one of great importance and the other which cannot be exhausted. The first is the utilization of sewage and garbage as fertilizer and the return of wastes of all sorts to the soil rather than funneling all these invaluable agricultural materials constantly off our soils by sending them through the cities to be dumped into streams, lakes and the sea. Slowly, all too slowly, plants are being established in some of our great cities which reclaim glycerine, fats and oils from garbage and sewage and convert the residue into good fertilizer to be returned to the field to increase the production per acre of land everywhere.

At Freeport, Texas, the Dow Chemical Company, pushing further and further the processes by which magnesium was extracted during the war from sea water, is succeeding in the recovery of many other valuable elements. The sea, to be sure, is the source from which all of us came when the first fish went ashore to live and, as was pointed out earlier in this book, virtually and possibly all the elements of the sea are still necessary in the production of humans who are intelligent and healthy. The sea is the repository of all the minerals and elements which have been washed down into it since the beginning of time and if they can be recovered at a reasonable economic cost in great quantities, the prospects of making all possible agricultural land highly productive both in quantity and quality are, of course, immense. In the past the process of recovering elements directly from the sea has been economically prohibitive in so far as their use in agriculture was concerned, but today the prospect of recovering an incredible amount of potential fertility at a reasonable cost is within sight.

During the war, through necessity, the United States was forced to produce magnesium needed in war industries out of sea water. The process was expensive at first but necessary if the war was to be carried on, and it marked the beginning of other more complicated processes by which minerals and elements vital to food production and agriculture can be recovered from the sea at a price where they can be utilized on sound economic terms in agriculture of steadily increasing yields. This means, of course, that the supply of fertilizer from this source alone is inexhaustible.

Some of the Neo-Malthusians have taken the negative approach to the problem of food—that the solution is not to raise more food and improve distribution and destroy waste, but to limit the increase in populations. While the factors concerning distribution and production of food are difficult enough to solve, the imposition of birth control (if it could be imposed) or sterilization of whole segments of population is, indeed, not only more difficult but impossible. In a country such as Japan, where there was no religious prejudice or pressure against contraception, and where, indeed, it received considerable support from some elements of government, a campaign for contraception produced no perceptible reduction in the birth rate. Indeed, the fundamental reason for the war-like aggression of Japan was the necessity for obtaining more food, more raw materials (many of them, like cotton, of agricultural origin) and more markets to produce more money with which to import food to feed a steadily increasing population.

Universally, the greatest number of children are produced at the lowest economic level of income and living standards, and this in itself serves to operate against birth control being a solution in a roundabout way for an insufficient and a badly distributed food supply. There are many reasons for this—the listlessness and indifference of people living at low living standards and poor diets, the lack of plumbing facilities and the old poet's assertion that the pleasures of the poor are few and unvaried. But behind all this there lies as well the theory adumbrated long ago by Darwin that when animals and plants are placed under adverse conditions, their capacity for reproduction increases and their reproductive activity becomes violent and exaggerated in order that the species may survive, and there is no reason to suppose that mankind is favored by exception to this theory. It is a generally recognized law that as living standards and diets rise, the birth rate declines and eventually stabilizes itself. It is a law well known to livestock farmers that overfed animals and in particular those fed on too rich foods breed in an indifferent and spasmodic fashion. In other words, the Neo-Malthusians who advocate birth control as a solution are blithely overlooking difficulties of a biological and indeed of a cosmic nature.

On the other hand, the Neo-Malthusians who advocate sterilization are inviting difficulties equally great and insoluble. In the first place, it would require a vast army of educational bureaucrats innumerable years to explain to hundreds of millions of the earth's population the difference between sterilization and castration, and even after that difference was explained, it would require another army of bureaucrats to carry out a plan which is against the whole cosmic urge of reproduction and the eternal and passionate desire of the peasant and the working man for children to comfort and care for him in his old age. Probably the only willing candidates for the sterilization school would be a small dribble of male and female rakes looking for an opportunity to continue their pleasures without adding to the food problems of the world.

We have plenty of agricultural and scientific knowledge, the amount and variety of it is constantly increasing and it is today available to all the governments and peoples of the world. If what we already know were simply applied to all the agricultural land of the world and the problem of proper distribution were given more and better consideration, the world could feed itself even now and even better, on the whole, than it has done in the past. The means are at hand and available. The solution is up to man and his governments— whether he chooses to go on reproducing himself in constantly increasing numbers while he produces less and less food or whether he and his governments choose to settle down and produce those great quantities of food which it is potentially possible to produce.

All of the factors listed above support the belief of the writer that the world can feed itself at any time if it wants to badly enough. The limiting factors are the incapacity of mankind to organize itself, the limitations upon exchange in trade, raw materials and food among nations and the incapacity of mankind, except in the cases of a few notable leaders to rise above a certain political level. Some aspects of the problem could perhaps be called economic. All of them are difficult of solution but not impossible.

Perhaps it is that those who have the power to bring about the solution have not enough interest to do so and the rest, the poor, the ignorant and the starving do not possess the energy, the intelligence or the power to do so. After all, man is still, by his own record, not yet so far above the level of other creatures on this planet, however much he may vaunt his superiority. Nobody knows this better than the good farmer.

The only final solution is that of abundance, of distribution and of a rise in living standards and diet and employment which goes with abundance and which can be had if man wants it badly enough. The ones who want and need it most are the poor and the humble who starve first without ever achieving the goal of having had enough to eat for one day of their lives. The politicians are usually able to take care of themselves by buying votes with subsidies and support price plans, rather than finding markets and proper distribution.

A Philosophical Excursion

The agrarian class is the first in utility and ought to be the first in respect. The same artificial means which has been used to produce a competition in learning, may be equally successful in restoring agriculture to its primary dignity in the eyes of man. It is a science of the very first order. . . . Young men closing their academical education with this, as the crown of all other sciences, fascinated with its solid charms, and at the same time when they are to choose an occupation instead of crowding the other classes, would return to the farms of their fathers, their own or those of others, and replenish and invigorate a calling, now languishing under contempt and oppression.

—The Living Thoughts of Thomas Jefferson
(*presented by John Dewey*)

Epilogue: *A Philosophical Excursion*

AT MALABAR when the shadows grow longer across the Valley and each day the Big House falls earlier beneath the deep shadow of the low sandstone cliffs, we know that winter is closing in. On a still day when we hear the whistles of the big Diesels on the Pennsylvania Railroad six miles away we know that we shall have fine clear weather, and when the sound comes from the opposite direction from the Baltimore and Ohio, we know that there will be clouds and rain. We know the time by the flight overhead of the big planes going north and south, and some of the pilots know us so well that on summer nights they blink their lights in greeting as they pass through the clear, still sky overhead.

In the barns and the fields Al and Simon know their cows—a hundred and twenty of them—by name, and they know their dispositions and what they like or do not like, from Jean, the bossy old Guernsey who must be started homeward first on her way from pasture before the others will go properly, to Inez, the Holstein, smart and temperamental, who once struck up a feud with Mummy, the feed-room cat, and was observed on two occasions shaking Mummy as a dog might shake her, when the unfortunate cat came within reach.

As Philip, who lived with us until he grew up and went away on his own, once said, "The trouble with Malabar is that it's always characters—characters never people . . . even down to the ducks." That's true of most farm people and especially true of farm people in hill country where over each rise in the land, just beyond each patch of woods, there lies a new world. In the old days before automobiles and telephones, "character" developed in old age into eccentricity. Today the change is not so great, but the independence, the strength of opinion and willingness to fight for an opinion still remain. These are not regimented people herding at night into subways to return to a cave somewhere high up in a skyscraper, living as man was never meant to live.

For the young people a farm is a kind of Paradise. One never hears the whine of the city child, "Mama, what shall I do now?" On a farm no day is ever long enough for the young person to crowd into its meager twenty-four hours all there is to be done. That, too, is true of the good farmer himself. No day is long enough. There is fishing and swimming, explorations of the woods and the caves, trapping, messing about the big tractors, playing in the great mows, a hundred exciting things to do which each day are new and each day adventurous.

But most of all there is the earth and the animals through which one comes very close to eternity and to the secrets of the universe. Out of Gus, the Mallard duck, who comes up from the pond every evening to eat with the dogs, out of Stinker, the bull, with his wise eyes and placid disposition, out of all the dogs which run ahead leaping and barking and luring the small boys farther and farther into the fields, a child learns much, and most of all that warmth and love of Nature which is perhaps the greatest of all resources, not only because its variety and beauty is inexhaustible but because slowly it creates a sense of balance and of values, of philosophy and even of wise resignation to man's own significance which bring the great rewards of wisdom and understanding and tolerance. It is not by senseless accident that the vast majority of the great men and women of the nation and those who have built it have come from farms or hamlets.

There is in all the world no finer figure than a sturdy farmer standing, his feet well planted in the earth, looking over his rich fields and his beautiful shiny cattle. He has a security and an independence unknown to any other member of society, yet, unlike the trapper or the hunter, he is very much a part of society, perhaps its most important member. The sharp eyes with the crow's-feet circling them like small halos, the sunburned neck, the big strong hands, all tell their story of values and of living not only overlooked but unknown to far too many of those who live wholly in an industrial civilization where time clocks and machines rule man instead of man ruling them.

Nothing is more beautiful than the big farm kitchen. It has changed with the times. The refrigerator, the electric stove, the quick-freeze and the cold room have supplanted the cellar, the root storage and the great black old range with its tank of boiling water on the side. The woodpile is gone from outside the door and the horses no longer steam as they stand patiently while the farmer comes in for a cup of coffee and a cinnamon bun. We tell the time nowadays not by the whistle of the old steam locomotives but by the passage overhead of the big flying flagship. But the good smell is still there in the kitchen and the farmer's wife is the same at heart, although in these times she is not bent with rheumatism at forty from carrying water and wood and bending over a washboard. At forty she is likely to be spry and young and busy with her clubs and neighborhood activities—as young-looking as her eighteen-year-old daughter who is a leader in the 4-H Club. And her husband does not rise at daylight and come in weary and bent long after dark. He keeps city hours, but during the day his work is half fun, because the drudgery has gone out of it. He is out of doors with the smell of fresh-turned earth rising to him from the furrow, the sight of a darting cock pheasant rising before his eyes in a kind of brilliant hymn to the morning. He, too, is young and sturdy at middle age and able to go places with his boys, to fish and hunt with them and attend their meetings.

A lot of things have changed on the farm of today, but the essence of the farm and the open country remains the same. The freedom is unchanged and the sense of security and independence and the good rich food and the beauty that lies for the seeing eye on every side and, above all, that satisfaction, as great as that of Leonardo or Shakespeare or any other creative artist, in having made something great and beautiful out of nothing. The farmer may leave his stamp upon the whole of the landscape seen from his window, and it can be as great and beautiful a creation as Michelangelo's David, for the farmer who takes over a desolate farm, ruined by some evil and ignorant predecessor, and turns it into a Paradise of beauty and abundance is one of the greatest of artists.

Of course, I am talking about the good farmer, the real farmer, and not that category of men who remain on the land because circumstance dropped them there and who go on, hating their land, hating their work and their animals because they have never discovered that they do not belong there, that they have no right to carry in trust the greatest of all gifts Nature can bring to man—a piece of good land, with the independence, the security, the excitement and even the splendor that go with it. The good farmer, working with Nature rather than fighting or trying to outwit her, may have what he wants of those treasures which are the only real ones and the ones by which man lives—his family, his power to create and construct the understanding of his relationship to the universe, and the deep, religious, humble sense of his own insignificance in God's creation.

The good farmer of today can have all the good things that his father knew and many that his father never knew, for in the modern world he lives with all the comforts of a luxurious city house plus countless beauties and rewards forever unknown to the city dweller. More than any other member of our society—indeed, perhaps alone in our society—the farmer has learned how to use machinery to serve him rather than his serving machinery. That is a very great secret indeed and one which the other members of our society need desperately to learn.

Index